COLLOID AND SURFACE PROPERTIES OF CLAYS AND RELATED MINERALS

SURFACTANT SCIENCE SERIES

1. Nonionic Surfactants, *edited by Martin J. Schick* (see also Volumes 19, 23, and 60)
2. Solvent Properties of Surfactant Solutions, *edited by Kozo Shinoda* (see Volume 55)
3. Surfactant Biodegradation, *R. D. Swisher* (see Volume 18)
4. Cationic Surfactants, *edited by Eric Jungermann* (see also Volumes 34, 37, and 53)
5. Detergency: Theory and Test Methods (in three parts), *edited by W. G. Cutler and R. C. Davis* (see also Volume 20)
6. Emulsions and Emulsion Technology (in three parts), *edited by Kenneth J. Lissant*
7. Anionic Surfactants (in two parts), *edited by Warner M. Linfield* (see Volume 56)
8. Anionic Surfactants: Chemical Analysis, *edited by John Cross*
9. Stabilization of Colloidal Dispersions by Polymer Adsorption, *Tatsuo Sato and Richard Ruch*
10. Anionic Surfactants: Biochemistry, Toxicology, Dermatology, *edited by Christian Gloxhuber* (see Volume 43)
11. Anionic Surfactants: Physical Chemistry of Surfactant Action, *edited by E. H. Lucassen-Reynders*
12. Amphoteric Surfactants, *edited by B. R. Bluestein and Clifford L. Hilton* (see Volume 59)
13. Demulsification: Industrial Applications, *Kenneth J. Lissant*
14. Surfactants in Textile Processing, *Arved Datyner*
15. Electrical Phenomena at Interfaces: Fundamentals, Measurements, and Applications, *edited by Ayao Kitahara and Akira Watanabe*
16. Surfactants in Cosmetics, *edited by Martin M. Rieger* (see Volume 68)
17. Interfacial Phenomena: Equilibrium and Dynamic Effects, *Clarence A. Miller and P. Neogi*
18. Surfactant Biodegradation: Second Edition, Revised and Expanded, *R. D. Swisher*
19. Nonionic Surfactants: Chemical Analysis, *edited by John Cross*
20. Detergency: Theory and Technology, *edited by W. Gale Cutler and Erik Kissa*
21. Interfacial Phenomena in Apolar Media, *edited by Hans-Friedrich Eicke and Geoffrey D. Parfitt*
22. Surfactant Solutions: New Methods of Investigation, *edited by Raoul Zana*
23. Nonionic Surfactants: Physical Chemistry, *edited by Martin J. Schick*
24. Microemulsion Systems, *edited by Henri L. Rosano and Marc Clausse*
25. Biosurfactants and Biotechnology, *edited by Naim Kosaric, W. L. Cairns, and Neil C. C. Gray*
26. Surfactants in Emerging Technologies, *edited by Milton J. Rosen*
27. Reagents in Mineral Technology, *edited by P. Somasundaran and Brij M. Moudgil*
28. Surfactants in Chemical/Process Engineering, *edited by Darsh T. Wasan, Martin E. Ginn, and Dinesh O. Shah*
29. Thin Liquid Films, *edited by I. B. Ivanov*
30. Microemulsions and Related Systems: Formulation, Solvency, and Physical Properties, *edited by Maurice Bourrel and Robert S. Schechter*
31. Crystallization and Polymorphism of Fats and Fatty Acids, *edited by Nissim Garti and Kiyotaka Sato*

32. Interfacial Phenomena in Coal Technology, *edited by Gregory D. Botsaris and Yuli M. Glazman*
33. Surfactant-Based Separation Processes, *edited by John F. Scamehorn and Jeffrey H. Harwell*
34. Cationic Surfactants: Organic Chemistry, *edited by James M. Richmond*
35. Alkylene Oxides and Their Polymers, *F. E. Bailey, Jr., and Joseph V. Koleske*
36. Interfacial Phenomena in Petroleum Recovery, *edited by Norman R. Morrow*
37. Cationic Surfactants: Physical Chemistry, *edited by Donn N. Rubingh and Paul M. Holland*
38. Kinetics and Catalysis in Microheterogeneous Systems, *edited by M. Grätzel and K. Kalyanasundaram*
39. Interfacial Phenomena in Biological Systems, *edited by Max Bender*
40. Analysis of Surfactants, *Thomas M. Schmitt* (see Volume 96)
41. Light Scattering by Liquid Surfaces and Complementary Techniques, *edited by Dominique Langevin*
42. Polymeric Surfactants, *Irja Piirma*
43. Anionic Surfactants: Biochemistry, Toxicology, Dermatology. Second Edition, Revised and Expanded, *edited by Christian Gloxhuber and Klaus Künstler*
44. Organized Solutions: Surfactants in Science and Technology, *edited by Stig E. Friberg and Björn Lindman*
45. Defoaming: Theory and Industrial Applications, *edited by P. R. Garrett*
46. Mixed Surfactant Systems, *edited by Keizo Ogino and Masahiko Abe*
47. Coagulation and Flocculation: Theory and Applications, *edited by Bohuslav Dobiáš*
48. Biosurfactants: Production • Properties • Applications, *edited by Naim Kosaric*
49. Wettability, *edited by John C. Berg*
50. Fluorinated Surfactants: Synthesis • Properties • Applications, *Erik Kissa*
51. Surface and Colloid Chemistry in Advanced Ceramics Processing, *edited by Robert J. Pugh and Lennart Bergström*
52. Technological Applications of Dispersions, *edited by Robert B. McKay*
53. Cationic Surfactants: Analytical and Biological Evaluation, *edited by John Cross and Edward J. Singer*
54. Surfactants in Agrochemicals, *Tharwat F. Tadros*
55. Solubilization in Surfactant Aggregates, *edited by Sherril D. Christian and John F. Scamehorn*
56. Anionic Surfactants: Organic Chemistry, *edited by Helmut W. Stache*
57. Foams: Theory, Measurements, and Applications, *edited by Robert K. Prud'homme and Saad A. Khan*
58. The Preparation of Dispersions in Liquids, *H. N. Stein*
59. Amphoteric Surfactants: Second Edition, *edited by Eric G. Lomax*
60. Nonionic Surfactants: Polyoxyalkylene Block Copolymers, *edited by Vaughn M. Nace*
61. Emulsions and Emulsion Stability, *edited by Johan Sjöblom*
62. Vesicles, *edited by Morton Rosoff*
63. Applied Surface Thermodynamics, *edited by A. W. Neumann and Jan K. Spelt*
64. Surfactants in Solution, *edited by Arun K. Chattopadhyay and K. L. Mittal*
65. Detergents in the Environment, *edited by Milan Johann Schwuger*

66. Industrial Applications of Microemulsions, *edited by Conxita Solans and Hironobu Kunieda*
67. Liquid Detergents, *edited by Kuo-Yann Lai*
68. Surfactants in Cosmetics: Second Edition, Revised and Expanded, *edited by Martin M. Rieger and Linda D. Rhein*
69. Enzymes in Detergency, *edited by Jan H. van Ee, Onno Misset, and Erik J. Baas*
70. Structure–Performance Relationships in Surfactants, *edited by Kunio Esumi and Minoru Ueno*
71. Powdered Detergents, *edited by Michael S. Showell*
72. Nonionic Surfactants: Organic Chemistry, *edited by Nico M. van Os*
73. Anionic Surfactants: Analytical Chemistry, Second Edition, Revised and Expanded, *edited by John Cross*
74. Novel Surfactants: Preparation, Applications, and Biodegradability, *edited by Krister Holmberg*
75. Biopolymers at Interfaces, *edited by Martin Malmsten*
76. Electrical Phenomena at Interfaces: Fundamentals, Measurements, and Applications, Second Edition, Revised and Expanded, *edited by Hiroyuki Ohshima and Kunio Furusawa*
77. Polymer-Surfactant Systems, *edited by Jan C. T. Kwak*
78. Surfaces of Nanoparticles and Porous Materials, *edited by James A. Schwarz and Cristian I. Contescu*
79. Surface Chemistry and Electrochemistry of Membranes, *edited by Torben Smith Sørensen*
80. Interfacial Phenomena in Chromatography, *edited by Emile Pefferkorn*
81. Solid–Liquid Dispersions, *Bohuslav Dobiáš, Xueping Qiu, and Wolfgang von Rybinski*
82. Handbook of Detergents, *editor in chief: Uri Zoller*
    Part A: Properties, *edited by Guy Broze*
83. Modern Characterization Methods of Surfactant Systems, *edited by Bernard P. Binks*
84. Dispersions: Characterization, Testing, and Measurement, *Erik Kissa*
85. Interfacial Forces and Fields: Theory and Applications, *edited by Jyh-Ping Hsu*
86. Silicone Surfactants, *edited by Randal M. Hill*
87. Surface Characterization Methods: Principles, Techniques, and Applications, *edited by Andrew J. Milling*
88. Interfacial Dynamics, *edited by Nikola Kallay*
89. Computational Methods in Surface and Colloid Science, *edited by Małgorzata Borówko*
90. Adsorption on Silica Surfaces, *edited by Eugène Papirer*
91. Nonionic Surfactants: Alkyl Polyglucosides, *edited by Dieter Balzer and Harald Lüders*
92. Fine Particles: Synthesis, Characterization, and Mechanisms of Growth, *edited by Tadao Sugimoto*
93. Thermal Behavior of Dispersed Systems, *edited by Nissim Garti*
94. Surface Characteristics of Fibers and Textiles, *edited by Christopher M. Pastore and Paul Kiekens*
95. Liquid Interfaces in Chemical, Biological, and Pharmaceutical Applications, *edited by Alexander G. Volkov*

96. Analysis of Surfactants: Second Edition, Revised and Expanded, *Thomas M. Schmitt*
97. Fluorinated Surfactants and Repellents: Second Edition, Revised and Expanded, *Erik Kissa*
98. Detergency of Specialty Surfactants, *edited by Floyd E. Friedli*
99. Physical Chemistry of Polyelectrolytes, *edited by Tsetska Radeva*
100. Reactions and Synthesis in Surfactant Systems, *edited by John Texter*
101. Protein-Based Surfactants: Synthesis, Physicochemical Properties, and Applications, *edited by Ifendu A. Nnanna and Jiding Xia*
102. Chemical Properties of Material Surfaces, *Marek Kosmulski*
103. Oxide Surfaces, *edited by James A. Wingrave*
104. Polymers in Particulate Systems: Properties and Applications, *edited by Vincent A. Hackley, P. Somasundaran, and Jennifer A. Lewis*
105. Colloid and Surface Properties of Clays and Related Minerals, *Rossman F. Giese and Carel J. van Oss*
106. Interfacial Electrokinetics and Electrophoresis, *edited by Ángel V. Delgado*
107. Adsorption: Theory, Modeling, and Analysis, *edited by József Tóth*

ADDITIONAL VOLUMES IN PREPARATION

Adsorption and Aggregation of Surfactants in Solution, *edited by K. L. Mittal and Dinesh O. Shah*

Biopolymers at Interfaces: Second Edition, Revised and Expanded, *edited by Martin Malmsten*

COLLOID AND SURFACE PROPERTIES OF CLAYS AND RELATED MINERALS

Rossman F. Giese
Carel J. van Oss

University at Buffalo
State University of New York
Buffalo, New York

MARCEL DEKKER, INC. NEW YORK • BASEL

ISBN: 0-8247-9527-X

This book is printed on acid-free paper.

Headquarters
Marcel Dekker, Inc.
270 Madison Avenue, New York, NY 10016
tel: 212-696-9000; fax: 212-685-4540

Eastern Hemisphere Distribution
Marcel Dekker AG
Hutgasse 4, Postfach 812, CH-4001 Basel, Switzerland
tel: 41-61-260-6300; fax: 41-61-260-6333

World Wide Web
http://www.dekker.com

The publisher offers discounts on this book when ordered in bulk quantities. For more information, write to Special Sales/Professional Marketing at the headquarters address above.

Current printing (last digit):
10 9 8 7 6 5 4 3 2 1

PRINTED IN THE UNITED STATES OF AMERICA

To former students whose contribution to this work was fundamental: Dr. Z. Li, K. Murphy, Dr. W. Wu and Dr. J. Norris. We especially acknowledge Dr. P. M. Costanzo, who encouraged us in this effort.

Preface

Much has happened in the field of interfacial interactions among clay and other mineral particles surfaces since the publication of van Olphen's book in 1977 (see also van Olphen and Fripiat (1979)). Up to the middle 1980s one still habitually only considered van der Waals attractions and electrostatic repulsions as acting between such particles, even when immersed in a polar liquid such as water. The stability of aqueous suspensions of low electric charge particles was loosely ascribed to "steric" interactions, which however cannot be quantitatively expressed, e.g., in SI units. Even among the three kinds of van der Waals forces: dispersion (fluctuating dipole induced dipole), induction (dipole induced dipole) and orientation (dipole dipole) interactions, respectively alluded to as van der Waals-London, or dispersion forces, van der Waals-Debye, and van der Waals-Keesom forces, it remained difficult to decide which of these interactions was genuinely apolar. It was therefore even more uncertain how to make a sharp delineation between apolar forces (usually alluding to dipersion forces only) and polar forces (including hydrogen-bonding interactions). It was Dr. Manoj Chaudhury (1984), who first applied Lifshitzs' theory to macroscopic-scale colloidal interactions, which allowed him to establish for the first time a clear-cut distinction between apolar, or Lifshitz-van der Waals (LW) and polar, or electron-acceptor/electron-donor or Lewis acid-base (AB) interactions. Once this distinction had become obvious, it fairly quickly permitted the elaboration of a surface-thermodynamic approach comprising both apolar (LW) and polar (AB) interactions (van Oss *et al.*, 1987) and also includes electrostatic (EL) interactions (van Oss *et al.*, 1988).

Whilst for EL contributions the measurement methodology is an electrokinetic one (usually via micro-electrophoresis), for LW and AB interactions together the method of choice remains the determination of contact angles with drops of a small number of appropriate, relatively high-energy liquids, apolar (e.g., diiodomethane) as well as polar (water, glycerol, formamide), deposited on a solid surface. This still remains the preferred approach, as it is the only one that allows the analysis of the surface properties of condensed-phase materials at their exact surface (not 1, 5, 10 or more nm below it). This is particularly important because, e.g., solid and liquid materials interact with one another precisely via their exact surfaces. Contact angle measurements are of course easy to perform on, e.g., flat surfaces of large mineral crystals. With small clay or other mineral particles, however, this is a somewhat different matter. When one has particles of a swelling clay (e.g., smectite particles) it is relatively easy to deposit such clay particles by sedimentation from an aqueous suspension onto a smooth surface of plastic or glass, whereupon after drying one obtains perfectly smooth surfaces (Giese *et al.*, 1990; van Oss *et al.*, 1990) with which contact angle measurements are quite easily performed. A difficulty arises however with non-swelling clays and with practically all other mineral particles. These can of course quite easily be deposited on a smooth flat surface but by their very granular nature they form, upon drying, a rough surface, usually with a radius of curvature of roughness of 1 or more μm. Unfortunately, surfaces exhibiting a degree of roughness greater than 1 μm, give rise to strongly exaggerated contact angles, so that direct contact angle analysis cannot be used in these cases (Chaudhury, 1984). Again, encouraged by Chaudhury (personal communication, ca. 1989), the thin layer wicking technique was developed by a number of us (Costanzo *et al.*, 1991; van Oss *et al.*, 1992), with the help of Karen Murphy and Drs. Janice Norris, Z. Li and Wenju Wu. This approach made it possible for the first time to measure the surface properties of many of the particles listed in Chapter 9. Only swelling clays allow the determination of their surface properties by direct contact angle measurement (see also Chapter 9), but wicking cannot be done with swelling clays because during the capillary rise measurements, involved in wicking, the liquid migration which should occur in one direction only, is severely diminished in that one direction, when liquid uptake occurs at the same time in two other directions, due to swelling.

There was hitherto no absolute certainty that contact angle measurements via contact angles and thin layer wicking really afforded measurements of the same contact angle values. However, it was subsequently found that the use of non-swelling monosized cubical synthetic hematite (Fe_2O_3) particles of about 0.6 μm, permitted the deposition of exceedingly smooth layers of particles onto glass surfaces. These could be utilized for thin layer wicking *as well as* for direct contact angle measurements and thus furnished the experimental proof that the same contact angles would result from either technique (Costanzo *et al.*, 1995).

Interfacial, non-covalent interaction energies between various identical, or different, condensed-phase materials, taking place in water, are here alluded to as ΔG^{IF}_{1w1} and ΔG^{IF}_{1w2}, respectively (see Tables 1 and 2). These represent either hydrophobic attractions (when $\Delta G^{IF} < 0$) or hydrophilic repulsions (when $\Delta G^{IF} > 0$). Hydrophobic attractions are often alluded to as caused by the "Hydrophobic Effect", whilst hydrophilic repulsions correspond to what has been called "Hydration Pressure". Both free energies of hydrophobic attraction and hydrophilic repulsion can be obtained from the measurements of surface properties of condensed-phase materials, and expressed quantitatively in SI units. Both types of interaction are relatively long-range: their interaction energies decay as a function of distance, ℓ, as $\exp[(\ell_o - \ell)/\lambda]$, where ℓ_o is the minimum equilibrium distance between two surfaces ($\ell_o = 0.157$ nm), and λ is the characteristic length for water ($\lambda \approx 1.0$ nm). Hydrophobic attraction is due to the Lewis acid-base (AB) free energy of *cohesion* between the water molecules (w), whereas hydrophilic repulsion is a consequence of the (AB) free energy of *attraction* between the water molecules (w) electron-acceptor sites (i.e., their H-atoms) and the electron-donor sites on the immersed hydrophilic materials or molecules (1 and/or 2). When the free energy of hydrophilic repulsion is greater than the underlying free energy of hydrophobic attraction (the latter, by its very nature is always present), a net repulsion ensues. Compared with AB interactions, occurring in water, Lifshitz-van der Waals (LW) interactions provide only a minor contribution to the interfacial free energies, ΔG^{IF}_{1w1} or ΔG^{IF}_{1w2}, as do electrostatic interactions (EL), which however for practical reasons are considered separately from interfacial (IF) interactions.

Most of the data generated on the surface properties of clays and other mineral particles, in the decade from 1988 to 1998 are discussed in the various Chapters of this book, especially in Chapter 9. A summary of some of the more important general applications of the above-described approach, treating polar interactions among clays and other mineral particles in aqueous media, is outlined in Tables 1 and 2.

The material presented in this book is designed to be of use to a wide variety of people, often with very different backgrounds. In order to provide a reasonably self contained volume, there are a number of sections which are

Phenomenon or Method	Mechanism	Utilizations	Chapter-Section
Stability of aqueous particle suspensions, as determined via the extended DLVO approach	Hydrophobic (AB) repulsion vs. LW attraction; see equations in Table 8.4	Determination of suspension stability, e.g., for the prediction of movements of soils, or of volcanic ash deposits; flocculation	8.8, 8.9
Connection between ΔG^{IF}_{1w1} of 1 with respect to solvent (e.g., water) and the solubility of 1	see Eqn. (8.7)	Computation of solubility, or of critical micelle concentration from ΔG^{IF}_{1w1}, or vice-versa	8.10
Swelling of clay minerals	Hydrophilic (AB) repulsion	formation of impermeable environmental barriers for containment of toxic or radioactive wastes	8.13, 8.15

Table 1: *Applications of quantitatively expressed* ΔG^{IF}_{1w1} *values, i.e., the free energies of interfacial, non-covalent interactions between identical particles or molecules, 1, immersed in water, w.*

Phenomenon or Method	Mechanisms	Utilizations	Chapter-Section
Adsorption and adhesion	Hydrophobic (AB) attraction; for dissociation, followed by hydrophilic (AB) repulsion	Liquid chromatography methods (e.g., reversed phase liquid chromatography), other separation and purification methods; removal of toxic spills; hydrocarbon recovery, various utilizations of zeolites; additives of clays in the paper industry; flotation	2, 8.11, 8.12,10

Table 2: *Applications of quantitatively expressed* ΔG^{IF}_{1w2} *values, i.e., the free energies of interfacial, non-covalent interactions between different particles and/or molecules, 1 and 2, immersed in water, w.*

intended as background material. For example, Chapters 3 and 4 provide an introduction to clay mineralogy and common minerals in general. Clearly, for a reader with a background in mineralogy, these chapters will be of marginal interest if any. The book also provides a summary of modern colloid chemistry (Chapters 5–8). More detail can be found in the book by van Oss (van Oss, 1994). The surface thermodynamic data on minerals are presented in Chapter 9. Clearly, not all minerals have been examined. The study of the surface chemistry of minerals is still young, so there are a number of minerals and mineral groups which have not yet been examined. It is the hope of the authors that the present volume will stimulate others to expand our knowledge.

The pictorial representations of mineral structures were created using the software package ATOMS, version 3.2 by Eric Dowty. P. Avery prepared many of the mineral samples for the measurements reported in Chapter 9.

Rossman F. Giese

Carel J. van Oss

Contents

COLLOID AND SURFACE PROPERTIES OF CLAYS AND RELATED MINERALS

1

Introduction

1.1 Importance of Clay Minerals

The earth is a large and complex object. It is differentiated according to the densities of the minerals which compose the planet. The density differences along with the abundances of the elements making up the earth ultimately lead to an accumulation of iron in the center (the core) of the earth, surrounded by a thick layer of silicate minerals relatively rich in iron and magnesium (the mantle), overlain by a thin layer of silicate minerals with less iron and magnesium and more silicon and aluminum (the crust). All of these materials, if placed on a laboratory table would be identified by a competent geologist as some sort of igneous or metamorphic rock, that is the product of heat along with pressure. Rocks at the surface of the crust are modified by contact with a very corrosive chemical: water in equilibrium with dissolved carbon dioxide to form carbonic acid. This weak acid coupled with an active atmosphere and moderate temperatures leads to the chemical and physical breakdown of a wide variety of rock types.

Thus, a mineral (part of a rock) that formed and was at equilibrium under conditions of elevated temperature and pressure is unstable at earth surface conditions. The acidic water, possibly aided by temperature fluctuations, attacks the mineral and a series of chemical reactions ensue producing a series of new minerals which are at equilibrium under the new conditions. If there is sufficient time, then ultimately the end of the search for thermodynamic equilibrium produces clay minerals with lesser quantities of other colloidal materials. This is the process known as weathering and is the major source of sediments and soils at the earth's surface on the continents and blanketing much of the oceans. Since the clay minerals form at low temperatures (in a geological sense), reaction rates are slow and crystals of these new phases form slowly and imperfectly resulting in very small particle sizes, far smaller than would result from mechanical abrasion of larger crystals.

We share the surface of the earth with these sediments, sedimentary rocks, and soils and our lives benefit greatly from the existence of these fine-grained materials. If weathering did not take place, the continents would be barren and unproductive places. The clay minerals form something like one third of the sediments (clay rich muds and silts), sedimentary rocks (principally shale) and soils.

Clearly a material that is so common at the surface of the earth would be of geological importance. In addition, clay minerals are of considerable technological importance as, for example, raw materials for ceramic ware, additives to a range of products including paints, inks and rubber. At one time, clay minerals were widely used as catalysts, but much of this market has been taken over by zeolites. The fertility of soils is largely due to the presence of clay minerals which contribute an ability to retain water and to exchange a variety of cationic species. These topics are discussed in some detail in Chapters 3 and 8.

The utility of clay minerals is strongly linked to their interfacial properties, especially with water. Thus, the ability to fabricate a complexly shaped ceramic body depends on the ability of the fine-grained raw materials, clay minerals, quartz and feldspar principally, to be formed while wet into the desired shape and retain that shape while drying. Without the plasticity of the wet clay, the shaping would not be possible. In fact, a clay is a clay because it is plastic when wet with an appropriate quantity of water. The plasticity is a result of complex interactions between the water and the surfaces of the constituents of the clay, principally the clay minerals.

The ability of clay minerals to participate in ion exchange reactions is a mechanism for the release of transition elements, alkali and alkaline earth metals to plant roots. Inorganic cations are not the only exchangeable entities; a wide variety of organic cations also undergo exchange leading to the conversion of the clay surface from neutral or hydrophilic to hydrophobic. A fine-grained hydrophobic material has very attractive properties as a barrier for contaminated soils or dump sites, in that it has a *large surface area* coupled with an *attraction* for hydrophobic organic compounds such as halogenated aromatic molecules. This is an area of intense experimental and commercial interest at present. Untreated clay minerals are frequently transported by rivers and streams. The transport is easy because of the small grain size of the clays, typically of the order of a few μm or less so that average-sized clay particles are suspended in stagnant water. However, the usual turbulence of natural flowing water can keep these particles in almost indefinite suspension. When conditions change, as will happen if the chemistry of the water is modified by, for example, the entry of river water into the ocean (which is richer in electrolyte) can cause the clay mineral particles to flocculate upon encountering the higher salt content of seawater. Again, this is the result of competing forces between the clay particles and water molecule interacting with clay particles, and the interactions of water molecules with each other. The traditional explanation of flocculation rests on the electrostatic interactions of clay particles as modified by the electrolyte. The real situation is much more complex and more interesting and will be discussed later (Chapter 8).

2

Applications of Clays and Clay Minerals

The uses of clay minerals and other colloidal materials are numerous. In this chapter, we review the major uses and try to indicate how the surface and interfacial properties of these materials influence the utility of the material in question. Further, we attempt to indicate how knowledge of the colloidal properties of these materials can be modified or how the surface properties might be better utilized.

2.1 Ceramics and Related Clay Products

By far the largest use of clay is for the manufacture of building, paving and facing bricks and other structural ceramic ware, followed by refractories, whiteware and porcelain, in that order (Clews, 1971).

2.1.1 Bricks and other structural ceramic ware

For building, facing and paving bricks, one generally uses mostly unpurified mixtures of clay, sand and other soil constituents, which are found as closely as possible to the place of manufacture in order to reduce transport costs. Bricks, and other items in this category are generally pre-shaped, air-dried (or dried at relatively low heat), to ensure proper pre-shrinkage, before firing in kilns, at a fairly low temperature of only slightly more than 1000°C. Clays used for brick manufacture can contain a significant proportion of swelling clays which, upon firing tend to cause considerable shrinkage, which is one of the reasons for applying pre-firing drying, and firing at a very slow rate. Also helpful in reducing shrinkage is avoidance of the use of very small hydrophilic clay particles so as to reduce the amount of water of hydration present per unit weight (van Oss and van Oss,1956).

2.1.2 Refractories

Refractories are ceramics, including bricks destined for use in ovens, electric insulators, etc., which must withstand rather high temperatures. To that effect refractory ceramics may not contain compounds that melt at the temperatures at which they are destined to be used. In particular Fe-containing compounds must be avoided. The major clay varieties used for refractories are kaolinite, quartz and illite. Here also the more finely divided varieties of illites in particular, are best avoided (van Oss and van Oss, 1956). Firing of refractories requires temperatures of the order of 1500-1600°C.

2.1.3 Earthenware

Earthenware, faience, or whiteware, is glazed pottery, i.e., relatively coarse, porous ceramic, glazed for decorative purposes and for impermeability (e.g., waterproofing). It includes tiles, tableware, mugs, ceramic pots and pans, etc. Earthenware can be dried much more quickly than bricks (see above), owing to its greater thinness. Firing temperatures usually are between 1000° and 1300°C. An important ingredient for earthenware (e.g., whiteware; see Clews (1971)) is ball clay, i.e., clay that contains organic matter which lends it an appreciable degree of plasticity: a desirable quality to facilitate forming or molding. Ball clay is somewhat colored, the darker the color, the greater the proportion of organic matter. The plasticity of ball clay is also partly due to its very small particle size (sub-micron) (Clews, 1971). Next to ball clay, earthenware clay would contain kaolinite, and quartz. The particle distribution and configuration of the unglazed part of earthenware tends to be random and fairly coarse, as can be readily seen on broken edges of pottery, plates or tiles of this category.

2.1.4 Porcelain

Porcelain is much finer grained than earthenware; it is impermeable and often somewhat translucent (Clews, 1971). An important proportion (i.e., about half, or somewhat more than half) of its constituent clay is made up of kaolinite (or china clay), with lesser admixtures of feldspar, and quartz (Clews, 1971). Porcelain ceramic ware has been made in China for many centuries, and in Europe (especially in France, England and Germany), since the end of the seventeenth Century.

Clews (1971) illustrates the compressive strength of porcelain with a photograph of an eight-wheel beer tank truck resting on eight bone china coffee cups, which is largely a consequence of the unusual particle configuration of colloidal kaolinite suspensions in aqueous media. Whereas virtually all other aqueous suspensions of clay particles owe their suspension stability to their mutual repulsion as a consequence of their general negative overall electric charge and their largely monopolar and strong electron-donicity (Giese *et al.*, 1996), kaolinite suspensions manifest a different and more complex colloidal behavior. Conventional colloids, such as for example, montmorillonites, typically form the most stable suspensions in water at the lowest salt concentrations (van Oss *et al.*, 1990a), because at the lowest ionic strengths, the electrostatic repulsion energies between charged particles are at a maximum (Verwey and Overbeek, 1948, 1999). Concomitantly with a strong electrostatic repulsion the particles also undergo a pronounced Lewis acid-base repulsion (Wu *et al.*, 1994a); see also Chapter 8, Sections 8.4, 8.8 and 8.9. Now, kaolinite particles, suspended in water reach a maximum of suspension stability at *high salt* concentrations and they become unstable (i.e., they flocculate) at *low ionic strength* (Schofield and Samson, 1954). This is because kaolinite particles, which like many clay particles, actually are small, flat platelets, have a negative charge on their flat surfaces (giving them an overall average negative charge), but they have in addition, positively charged edges. At high ionic strength both positive and negative charges are attenuated but the residual negative charge remains sufficiently dominant for overall repulsion to prevail, thus allowing suspension stability. At low salt concentrations, however, both the flat-negative and the edge-positive charges are enhanced, so that plus-minus attractions between plates and edges take over, which gives rise to a flocculation of the platelets in an edge-flat-plate, or "house of cards" conformation (van Olphen, 1977). With kaolinite particles, however, the edges are much thicker than with any other clay particles (Caillère *et al.*, 1982) which explains their opposite behavior of stability vis-à-vis low and high ionic strength media; see also Section 8.16. Upon drying and firing of largely kaolinite-based ceramics, i.e., porcelain, this "house of cards" conformation can therefore persist to a significant degree, which results in structures of marked homogeneity, low porosity and high (especially compressive) strength, even though in porcelain a high proportion of fairly large platey particles (of about 6 to 10 μm) is used (Clews, 1971). Porcelain, and also bone china, is fired at about 1200° to 1250°C.

A variety of porcelain which is mainly popular in England, is alluded to as bone china. In bone china, about 1/3 of the ingredients consists of calcined bone. When mixed with kaolinite and ground (Cornish) stone (a weathered feldspar), a eutectic forms, so that upon firing (at about 1250°C) partial melting occurs, which causes translucency, further diminishes the porosity and lends it exceptionally high compressive strength (Clews, 1971).

An important specialized utilization of porcelain is in electrical porcelain, as in, e.g., electrical insulators. Here high dielectric strength, which is concomitant with low porosity, is an important property of electrical porcelain. Steatite (soapstone, a massive talc-rich rock) can also be an ingredient in (especially low tension) electrical porcelain.

2.1.5 Pencil leads

Pencil leads are made of graphite and very fine particle, plastic ball clay. The greater the relative amount of clay, the harder the pencil lead will be. The ball clay must be very pure and especially devoid of gritty impurity particles (Smith, 1971).

2.2 Clay as Filler Material

The most important use of clay as filler material (apart from agricultural uses; see below) is in the paper industry where the clay particles fill the voids between the cellulose fibers lending an improved body to the paper, increasing the brightness (i.e., reflectivity) and whiteness of the paper and preventing bleed through from printing on one side of the paper to the other side (Warren, 1973). The clay used is fine china clay (kaolinite), devoid of all coarse or gritty material. Also used as filler in paper-making are $CaSO_4$ (anhydrite), $CaCO_3$ (calcite), $MgSiO_3$ (enstatite), $BaSO_4$ (barite) and TiO_2 (rutile). For coating, mainly materials such as $BaSO_4$ are used (Warren, 1973).

Another use of clay as filler materials is in the paint industry, as "extenders." In addition to $BaSO_4$, $CaCO_3$, $CaMg(CO_3)_2$ (dolomite) and SiO_2 (quartz or flint), kaolinite, smectites, talc, mica and up to a decade or so ago, asbestos (Warren, 1973) are used as extenders in paints. Smectites are especially helpful in improving the suspension stability (i.e., preventing the sedimentation of solids) of paint (Scott and Witney, 1972).

2.3 Agricultural Applications

Clay plays a pervasive role in modulating the quality of soils destined for agriculture. Clay particles provide particle smallness, and thus a tremendously large surface area per gram of soil, often of the order of 10 m^2/g, or more. Mixed with larger particles (e.g., silt or sand) a relatively low porosity is obtained (for instance, yielding only 40% interstitial space for air or water), while providing a strong capillary transport of water. The large specific surface area enhances the ion exchange effect of the clay minerals, thus helping the control of pH of the soil, as well as the supply of essential cations; see e.g., (Low, 1975).

One agricultural application of a naturally occurring optimal mixture of sand and clay is found in the Netherlands, close to and parallel with the sandy North Sea coast of Holland. That mixture, in a strip only a few miles wide and about 25 miles long, at the interface between coastal sand and inland clay, is uniquely suited for the cultivation of flower bulbs.

For ameliorating ion exchange, montmorillonites (smectites) are especially useful. With respect to potassium binding, kaolinites bind the least, and, e.g., mica and illite, bind the most K^+ (Low, 1975).

2.4 Clays as Adsorptive Materials

There are three modes by which clays and other materials can exert non-covalent adsorptive power on various molecules, from the liquid or gaseous state. These are: 1) physical, non-ionic adsorption onto the surfaces of finely divided materials, such as clay particles with large surface areas that are comprised in a small volume, 2) ion exchange, by electrostatic interaction and exchange, and 3) zeolitic action, by inclusion of small molecules in cavities or pores, and partial or complete exclusion of larger molecules by such small cavities.

2.4.1 Physical adsorption

Clays, especially smectites (in the guise of Fuller's earth), have been used since antiquity for the adsorptive removal of impurities or undesired color from wool (Robertson, 1986), oils, fats and waxes (Spaull, 1971); see also Section 2.5. The decoloring action of clays, and especially of smectite clay minerals, caused the latter to be frequently alluded to as "bleaching earth." The adsorptive properties of clays are also of use in petroleum refining, e.g., for the removal of gum-forming components, and for decoloring (Wall, 1972).

2.4.2 Ion exchange

For a cation-exchanging clay, e.g., a montmorillonite, M, one can write:

$$MA + B^+ \rightleftarrows MB + A^+ \qquad (2.1)$$

where A^+ and B^+ are cations. The montmorillonite can be used as a cation exchanger in any of the forms, MA, MB, etc. In the chemical industry, as well as for purposes of water-softening, very extensive use is made of ion exchange reactions. However, in most cases, one uses synthetic organic ion exchange materials, such as, typically, sulfonated polystyrene, as a cation exchanger. For water softening, i.e., the removal of Ca^{2+} from "hard" water one uses a cation exchanger, X, in the sodium form:

$$2XNa + Ca^{2+} \leftrightarrow XCa + 2Na^+ \qquad (2.2)$$

For agricultural applications mineral (clay) ion exchangers remain important (cf. Section 2.3). The cation exchanging properties of, e.g., montmorillonites are also useful for binding organic cations. One example of this is the binding of alkaloids (which are positively charged), and thus as an antidote for a variety of poisons, which often are of alkaloid nature (Robertson, 1986). For the relatively minor influence of different exchanged cations on the surface properties of various ion exchanged clays, see Table 9.1. The binding of cation exchanging montmorillonites to alkyl amines is important in the formation of bentones, which are used, e.g., in gel formation of oil and petroleum products (Spaull, 1971); see Section 2.7.

2.4.3 Zeolites as molecular sieves

Zeolites are natural or synthetic crystalline materials which can serve as molecular sieves with small, precisely defined cavities or pores. They can be used in the separation of gas molecules, according to molecular size. Spaull (1971) describes three classes of molecular sieve materials: 1) those with pore diameters between 4.9 and 5.6 Å, which exclude isoparaffins and aromatic molecules, but occlude n-paraffins and smaller molecules, 2) those with pore diameters between 4.0 and 4.9 Å, which exclude all paraffins except ethane, methane and oxygen and; 3) those with diameters between 3.8 and 4.0 Å, which occlude argon, nitrogen and smaller molecules. Molecular sieves can be used in the separation of a variety of gaseous and liquid phase mixtures, as well as in the role of supports for many chemicals, e.g., with the purpose of segregating toxic or badly smelling chemicals, and/or for the slow release of occluded chemicals (Spaull, 1971). Zeolites are also used as purification agents in petroleum refinery (Wall, 1972).

2.5 Washing Scouring and Felting: Fuller's Earth

Clays, especially smectites, significantly predate soaps as washing agents. In antiquity, clay were used for washing and bleaching textiles and unwoven wool. Intensive use of bleaching earth (mainly consisting of a Ca-smectite) caused wool fibers to cohere into thick strong layers of felt; hence the designation of Fuller's earth for these clays. The history of Fuller's earth from antiquity to the present, including its variegated uses over time, has been recorded by Robertson (1986) in an admirable and unique fashion, with extensive references. Salient modern applications of smectites are treated in Section 2.7.

2.6 Talc and Its Uses

Talc and pyrophyllite are the most hydrophobic of all clays, given their exceedingly low γ^{AB}-values, of 5.1-6.5 mJ/m^2 (Giese *et al.*, 1996); see also Chapter 9. (For an explanation of these symbols, see Chapter 5.) For thoroughly dried talc, the γ^{AB}-value is as low as 1.7 mJ/m^2 (van Oss *et al.*, 1997). In all cases, both the $\gamma^{\oplus}$ and $\gamma^{\ominus}$-values of talc are below 7 mJ/m^2. This observation is important, because very hydrophilic materials, when dry, can have a γ^{AB}-value of zero, but a $\gamma^{\ominus}$-value as high as 64 mJ/m^2; to be hydrophobic, a material must have a low γ^{AB}-value arising out of low values for both $\gamma^{\oplus}$ and $\gamma^{\ominus}$; cf. van Oss (1994). Both talc and pyrophyllite are 2:1 layer silicate minerals, without ionic substitution and are thus without a layer charge and interlayer cations. The external surfaces of mineral particles are largely 001 surfaces composed of oxygen atoms whose valencies have been completely satisfied in the sense of Pauling's electrostatic rule. Although the valencies of oxygen atoms are also completely satisfied in, e.g., polytheylene oxide (PEO), which is exceedingly hydrophilic as a consequence of the residual hydrogen-bonding capacity of its oxygen atoms, the difference is that in PEO the -CH_2-O-CH_2- bond is quite flexible, whereas in talc and pyrophyllite, the O atom's bond is extremely rigid. It is this rigidity of the oxygen atom in talc, with its completely satisfied bond that accounts for the very low (but not totally zero) electron-donicity, and thus limits the ability of the lone pair electrons of oxygen atoms at the external 001 surfaces that results in a weak Lewis basicity (low $\gamma^{\ominus}$-value) and concomitant hydrophobicity of talc. With a $\gamma^{\oplus}$ of zero for both PEO and talc, the $\gamma^{\ominus}$ of the former is 64 mJ/m^2, whilst for dry talc it is only 6.9 mJ/m^2 (van Oss *et al.*, 1997).

An interesting and different explanation of the hydrophobicity of talc has been proposed by Michot *et al.* (1990) being due, according to these authors to the strong affinity of gaseous nitrogen (presumably from the atmosphere) for the external surfaces of talc. However, the authors did not show that the affinity of nitrogen for talc is greater than that of other gases, nor did they show that the affinity of nitrogen for talc is greater than the affinity of nitrogen for other clay mineral surfaces. Finally, the authors did not explain why the adsorption of nitrogen (which after all represents some 78% of our atmosphere, by volume) to any other solid surface does not cause the pronounced hydrophobicity that is so noteable with talc, nor did they divulge what, if anything, makes nitrogen hydrophobic, especially when adsorbed on a solid substrate.

The uses of talc rely on its inherent softness and its hydrophobic nature, as shown by contact angle measurements, cf. Table 9.2 (Giese *et al.*, 1996; van Oss *et al.*, 1997). The flaky softness of powdered talc makes it ideal for many applications requiring non-hygroscopic, non-abrasive and non-sticking properties. It is therefore used as filler material in paint and also in rubber manufacture, as an anti-stick or dusting agent (Kuipers *et al.*, 1972). For the same reason talc powder is exceedingly widely used in cosmetics (Williams, 1972), as adjuncts to or major constituents of powders, as well as in pressed powders. Talc is also used as filler material in a variety of pharmaceutical applications. However, the application of talc where it can be in contact with mucous surfaces, is decreasing on account of the growing realization of it bio-hazards, see Section 10.5.

2.7 Smectites and Their Uses

2.7.1 Uses of hydrophilic smectites

The property of smectites as swelling clays when exposed to water, and their concomitant propensity to form gels when swollen, makes them uniquely appropriate for a variety of applications.

In drilling for petroleum, smectites are used in drilling muds (Robertson,

1986) using bentonites (a commercial name for a rock material rich in smectite), expecially in the sodium form. Gel-forming sodium bentonites cheaply and effectively prevent the deterioration of drilling channels through crumbling or quicksand formation. An added advantage of gel-forming smectites is their thixotropic behavior. This is the property of such gels to remain highly stable and very viscous when undisturbed, but to become liquid and to assume a low viscosity when mechanically disturbed. Thus the drilling channel, when left alone remains intact, once formed, whilst during the movement of the drill string locally creates a low-viscosity fluid which greatly facilitates the drilling, but only just as long as the drilling continues.

The same property of thixotropy is of advantage in the use of smectites in pharmaceutical preparations (Robertson, 1986), and in paints. In oil-based paints it is the organo-smectites, or bentones that are used to control viscosity. The advantage of the use of thixotropic smectites as additives in paints is that while painting with a brush or roller, the agitation brought about by the action liquefies the paint. This facilitates the act of painting, but as soon as the painting stops, gellification sets in, and the previously liquid paint becomes much more viscous and tends not to form drips or runs on the surface being painted.

Smectites are also important in foundry bonding (Robertson, 1986). Here molten metal is poured into molds formed from a mixture of sand, smectite or other clays, and water; the gellifying action of the smectite bonds the sand grains together and also gives a smooth surface to the mold.

The gel-forming properties of smectites also are used in forming impermeable barriers around land-fills or contaminated soils. However when such materials are heated under humid conditions as may occur in the close vicinity of high level radioactive waste, the contact of wet steam with the smectites can cause the mineral to change from hydrophilic to hydrophobic, which results in the loss of the gel-forming capacity and thus the low permeability of the clay liner will be lost (Couture, 1985; Bish *et al.*, 1999).

2.7.2 Bentones

Bentones or organo-clays are smectites (bentonites) in which the normal inorganic cation is replaced by an organic cation via a simple ion exchange reaction in an aqueous medium. The organic cations are typically quaternary ammonium cations (quats) such as hexadecyl trimethyl ammonium. In this manner the charged end of the quat is in contact with the clay surface near to the site of the layer charge while the alkyl tail is directed away from the clay surface, thus rendering the organo-clay fairly strongly hydrophobic

($\gamma^{\ominus}$-values from as low as 0.1 mJ/m^2 to barely below 28 mJ/m^2) depending largely on the alkyl chain length; see Table 9.3 and Giese *et al.* (1996).

In contrast, when talc, which is already rather hydrophobic ($\gamma^{\oplus} \approx \gamma^{\ominus} \approx$ 2.5 mJ/m^2), is treated with an amine (there is no ion exchange because of the lack of a layer charge on talc) such as octadecylamine, at fairly low coverage, by gentle heating of the amine mixed with talc, it can become virtually totally hydrophobic (Table 9.4, Giese *et al.* (1996)).

3

Clay Minerals

The terms "clay" and "clay mineral" are used in very different contexts. For example, a common interpretation of a clay substance is a material whose constituent particles are very small (i.e., "clay sized"). This is typically an engineering and sedimentological usage. Curiously, it is only very recently that clay mineralogists have agreed on meanings for the two terms at the head of this paragraph (Guggenheim and Martin, 1995). The term "clay" now refers to any material which exhibits a plastic behavior when mixed with water, while "clay mineral" refers to materials which have a layer structure or a structure substantially derived from or containing major features of such layer structures. Thus, clay minerals will be, e.g., kaolinite, illite, and smectite but not, e.g., micas, talc, and pyrophyllite which occur in much larger particle sizes. A complication inherent with clays is the occurrence of materials normally considered to be amorphous or having a considerable concentration of defects, such as stacking faults. Clay minerals typically form at low temperatures (by geological standards), at low pressures (ambient or slightly elevated) and in the presence of abundant water. Under these conditions, perfection in the organization of the crystal structure is unlikely. The details of the crystal structure of these materials is of great importance in understanding the physical and chemical properties of clays. This also is true of the highly disordered or amorphous materials where there still exists short range order.

3.1 Silicate Mineral Structures

The determination of the crystal structures of minerals began with the X-ray diffraction experiments of W. L. and W. H. Bragg beginning in 1914 (Bragg, W. L., 1914a,b; Bragg, L. and Claringbull, 1965). This led to an intensive study of the crystal structures of minerals by V. M. Goldschmidt and many others. We now know the structural details of all the common minerals as well as many of the rarer species. These studies show the arrangement of the atoms in the bulk material, but the surface and interfacial properties of solids are related to the arrangement of the atoms at and near the external surface of crystals. Our understanding of the surface structure of minerals is not nearly as refined as is our knowledge of the interior structure. The clay minerals are a happy exception, for reasons outlined below. More detailed information about the structures of minerals can be found in references such as Klein and Hurlbut (1993) and, for clay minerals, Moore and Reynolds (1997), Brindley and Brown (1980) and Bailey (1984, 1988b). We now have a great deal of information about the crystal structures of the ordered phyllosilicate minerals, and, in many instances this includes the hydrogen positions of hydroxyls and water molecules. The aim of this chapter is to describe the structures of the major minerals with sufficient detail so that a non-mineralogist can appreciate the complexity of these materials, especially as that complexity may relate to the surface properties. This involves judgment on the part of the authors; for example, it seems unlikely that polytypism, as a purely structural phenomenon, will have much influence on the surface properties of micas, so this subject is largely ignored, although the nomenclature of polytypism is explained. On the other hand, hydroxyl orientations may be important so some effort has been expended to include this information.

It is not the intent of this chapter or the next to describe all minerals but rather to concentrate on those minerals and minerals groups whose surface properties are of general interest, and also for which there are reliable surface thermodynamic measurements. For this latter reason, feldspars and pyroxenes, among others, are not treated. Further, since the description of clay minerals is frequently not very extensive (and frequently erroneous) in standard mineralogy texts, this aspect of mineralogy has been examined in some depth in this chapter.

Silicate minerals are oxides of silicon and a small number of elements from the first three columns of the periodic table and the transition elements. As such they closely mirror the abundance of the elements in the crust of the Earth (Table 3.1). Since the number of different elements which play a major role in the structure of silicate minerals is small, it is not surprising that the fundamental building blocks of these minerals, and many other non-silicate minerals, are few.

What follows is a very brief discussion of the crystal structures of silicate minerals and a few other geologically important minerals. The aim is to present the reader with an introduction to the subject, sufficient to allow

Element	Coordination Number	Polyhedron	Ionic Charge	Ionic Radius (Å)
Silicon	4	tetrahedron	+4	0.26
Aluminum	4	tetrahedron	+3	0.39
	6	octahedron	+3	0.54
Iron	6	octahedron	+2	0.78
	6	octahedron	+3	0.65
Calcium	6	octahedron	+2	1.00
	8	cube	+2	1.12
Sodium	6	octahedron	+1	1.02
	8	cube	+1	1.18
Potassium	6	octahedron	+1	1.38
	8	cube	+1	1.51
Magnesium	6	octahedron	+2	0.72
	8	cube	+2	0.89

Table 3.1: *The most commonly observed coordination polyhedra for the common elements in silicate structures in order of decreasing amount in the Earth's crust, omitting oxygen which is the most abundant element (from Klein and Hurlbut (1993)).*

him to follow the discussions of the surface properties of these materials. In order to emphasize the principles, much detail has been omitted. Such omissions can also be justified given our relatively primitive understanding of the structure of the surfaces of most minerals, that part which is of most importance to the surface thermodynamic properties.

3.1.1 The polyhedral paradigm

Pauling, drawing on the work of Goldschmidt and others, formulated the structural principles by which the majority of minerals were constructed. The basic building blocks are simple Platonic polyhedra, largely tetrahedra and octahedra, which represent the placement of oxygen atoms (the apices of the polyhedra) and the smaller cations (at the centers of the polyhedra). The number of oxygen atoms arranged about a cation is termed the coordination number (CN); the smaller the cation, the smaller the CN. Table 3.1 summarizes the geometrical constraints for the common elements in phyllosilicate silicate minerals. Figure 3.1 shows a tetrahedron and an octahedron in both aspects, i.e., as a polyhedron and as the arrangements of oxygen coordinating

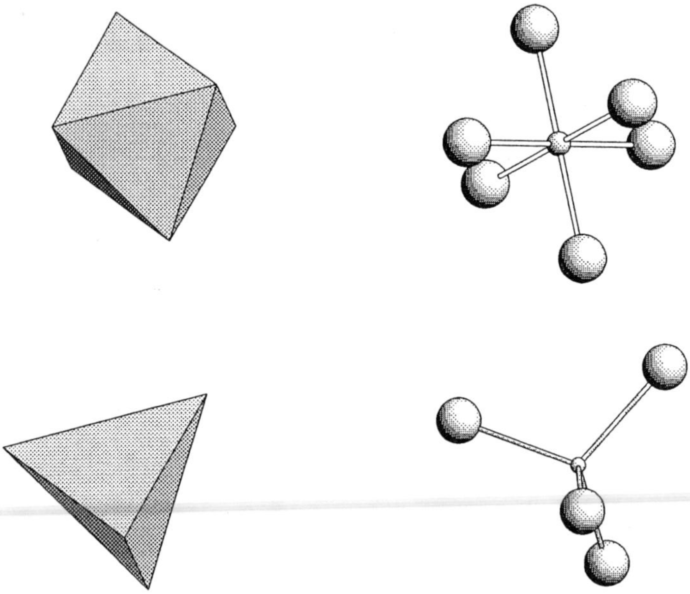

Figure 3.1: *Polyhedra (left) and the corresponding atomic arrangement (right) for the two principal polyhedra of silicate mineral structures, i.e., octahedra (upper) and tetrahedra (lower).*

the central cation. Note that while these drawings show perfect polyhedra, in real mineral structures, these are rarely perfectly regular. For example, that the edges of the polyhedra are almost always of slightly different lengths and the central cation may be displaced from the geometric center. For the purposes of this discussion such deviations are not important.

Such a view of crystal structures leads to a simplistic but nonetheless very useful concept of a silicate mineral, that is, a mineral is an arrangement of boxes in space (the coordination polyhedra), and we construct such a mineral by filling the boxes with appropriate cations. In a simple structure there might be only two kinds of boxes, representing tetrahedra and octahedra, appropriately linked together. Thus, we can change the chemical composition of a mineral by replacing all or part of the cations in one type of box by

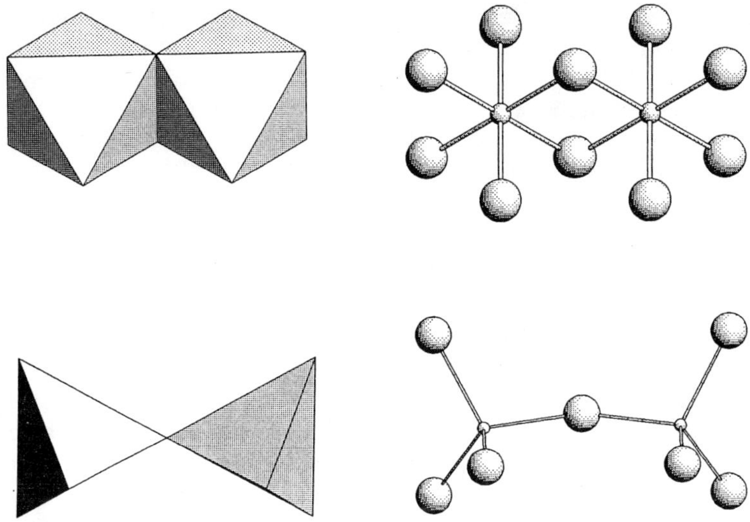

Figure 3.2: *The polyhedral (left) and equivalent atomic arrangement illustrating two common modes of sharing between polyhedra. The upper diagrams show two octahedra sharing an edge (two oxygen atoms) and the lower diagrams show two tetrahedra sharing a corner (a single oxygen atom).*

another kind of cation, such that size and valence considerations are not violated. Two divalent cations which are not very different in ionic radius, magnesium and iron, for example, can readily substitute for each other in an octahedral site.

3.1.2 Polymerization of polyhedra

Polyhedra polymerize to form inorganic crystal structures. The manner in which they do this is by sharing oxygen atoms. Geometrically, this is described as a sharing of edges and corners by adjacent polyhedra. Since the corners of coordination polyhedra are typically oxygen atoms, corner sharing simply means that a (single) shared oxygen atom is coordinating two neighboring cations. Similarly a shared edge represents two oxygen atoms which both belong to adjacent coordination polyhedra. These two aspects of mineral crystal structures are shown in Figure 3.2 for tetrahedra and octahedra.

Pauling pointed out the tendency for tetrahedra to only share corners while octahedra share corners and edges. Such sharing can be simple; for example, a structure which contains only isolated tetrahedra, i.e., tetrahedra which do not share any of their corners with other tetrahedra, or, at the other extreme, a structure in which each tetrahedron shares all four of its corners with adjacent tetrahedra. Mineralogists have adopted the view that the tetrahedra are of major importance in determining the physical and chemical properties of a silicate mineral, so that the manner in which the tetrahedra polymerize has been used as the basis for the classification of the silicate minerals.

3.2 Silicate Classification

Minerals are designated by a name and a chemical composition, a situation which may be confusing for non-mineralogists. For example, the silicate mineral olivine can have a composition which varies between the end members Mg_2SiO_4 and Fe_2SiO_4. The variation in chemical composition occurs by the simple interchange of Mg^{2+} and Fe^{2+} in an octahedral site, both cations having the same ionic charge and nearly the same ionic radius. One can indicate this variability in another way by writing the formula as $(Mg,Fe)_2SiO_4$, as a general description, or, if the actual composition is known, as, for example, $Mg_{0.37}Fe_{0.63}SiO_4$. The latter version is frequently referred to as a "structural formula" because it shows directly the chemical nature of each of the structural units in the mineral. In this case, the Mg and Fe occupy a single kind of octahedral site and the silicon occurs in a single tetrahedral site. It is usual to list the octahedral cations before the tetrahedral cations, and, where appropriate, cations in higher coordination sites precede both of these.

Olivine is a particularly simple case because the chemical variation principally involves a single octahedral site and not the tetrahedral cation, Si^{4+}. Other silicate minerals are more complex having a number of independent polyhedral sites and cationic substitutions in each. For these, the conversion of a chemical analysis into a structural formula is not a trivial exercise.

The silicate structure types are illustrated in Figure 3.3. The names which are traditionally given to each of these are indicated in the figure as are the chemical formulae for the tetrahedral part of the structures. For the latter, the tetrahedral cation is indicated by "Si" and the anion is taken to be oxygen, i.e., by "O." The charge (per cation) on the anion decreases as the degree of polymerization increases and ultimately becomes zero for silica, SiO_2. It should be noted that other cations frequently substitute for silicon in silicate mineral structures, and this substitution can change the charge on the anion. For example, substitution of aluminum (3+) for some of the silicon atoms in SiO_2 will generate a negative charge on the anion group. The existence of a negative charge then requires the addition of other cations which, for the most part, have a non-tetrahedral environment.

3.3 Structure of Phyllosilicates

As the name indicates, phyllosilicate (from the Greek, phyllon, a leaf, as in a leaf of a plant) minerals are layer structures. That is, the atomic arrangement is such that there are easily recognizable quasi two-dimensional fragments, strongly bonded internally, which are stacked one on top of the other with a much weaker bonding between the layers. Mechanically, the layers are more strongly bonded than the interlayers so that cleavage with always occur along the interlayer regions. This is the case for micas and vermiculites, both of which occur in large crystals. Such an arrangement creates a fundamental anisotropy; the layer is different and has different properties compared to the interlayer. The stronger and more specific the interlayer bond is, generally, the more regular and perfect will be the stacking. This is typically the situation encountered with the mica minerals. Conversely, the weaker and more diffuse the interlayer bonding, the less regular the stacking, the greater the number of stacking faults, and ultimately, one has a near-random stacking which is termed "turbostratic." Such is the case for a great many clay minerals, particularly the smectite minerals.

3.3.1 Layer types

There are several aspect of the overall crystal structure of phyllosilicates which are important in determining the properties of these materials and for their nomenclature. To begin with one distinguishes: a "plane" of atoms (e.g., the plane of oxygens forming the surface of a mica layer), a "sheet" as in the octahedral sheet of a mica layer, and finally a "layer." Of primary importance is the nature of the layer itself; the manner in which the layers are stacked is of secondary importance.

As was shown in Figure 3.3, there is a net negative charge on a sheet of tetrahedra as found in phyllosilicate minerals. This charge can be balanced by the addition of other cations, and, in phyllosilicates, this is done by the addition of a sheet of octahedra. Examination of the figure shows that the sheet of tetrahedra is composed of two parallel planes of oxygen atoms with the tetrahedral cations sandwiched between. Similarly, a sheet of octahedra

Silicate Unit	Ideal Formula	Silicate Class
	$(SiO_4)^{4-}$	Nesosilicate
	$(Si_2O_7)^{6-}$	Sorosilicate
	$(SiO_3)^{2-}$	Cyclosilicate
	$(SiO_3)^{2-}$	Inosilicate *single chain*
	$(SiO_4O_{11})^{6-}$	Inosilicate *double chain*

Figure 3.3: *The structural classification of the silicate minerals. Note that, with the exception of the tektosilicate structure, these are idealized models.*

$(Si_2O_5)^{2-}$ Phyllosilicate

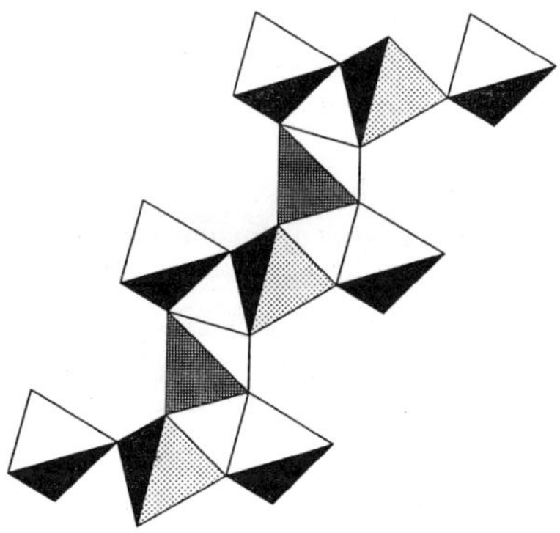

$(SiO_2)^0$ Tektosilicate

Figure 3.3. continued.

contains two parallel planes of oxygen atoms with the octahedral cations lying between. In an ideal tetrahedral sheet, as shown in Figure 3.3, the oxygen atoms are disposed on a simple hexagonal grid, with a number of atoms missing. The tetrahedra are all oriented with their apical oxygen atoms, the ones not being shared by other tetrahedra are oriented in the same direction. The opposite plane is the so-called "basal plane."

The spacing between the apical oxygen atoms is very similar to the spacing between oxygen atoms of the two planes forming a sheet of octahedra. Thus, the marriage of an octahedral sheet with a tetrahedral sheet is simply accomplished by having the tetrahedral apical oxygen atoms shared by the octahedral and tetrahedral sheets. The two sheets thus form a layer, the identity being given by the name "1:1", as in 1 octahedral layer and 1 tetrahedral layer. Such a layer is composed of 3 planes of oxygen atoms. The ionic charge on the basal oxygen atoms of the tetrahedral sheets is balanced, or largely, so depending on the tetrahedral cation charge, similarly, many of the oxygen atoms belonging to the middle plane of the layer and shared by both tetrahedra and octahedra are balanced, but the atoms of the third oxygen layer, having only octahedral cations as ligands, are not balanced and have, on average, a charge deficiency of 1 electron each. This difficulty is overcome when these oxygens become hydroxyls. Thus, in a 1:1 layer, there are two kinds of external surface; a purely oxygen plane, the basal tetrahedral oxygens, and a purely hydroxyl plane. Not all of the oxygen atoms of the central plane are shared between the tetrahedral and octahedral sheets; this arises because of the peculiar hexagonal arrangement of the tetrahedra. Those oxygens which do not have a neighboring tetrahedral cation form part of hydroxyl groups.

Figure 3.4a shows a projection of a 1:1 layer silicate oriented with the octahedral sheet uppermost. Note that the hydrogen atoms are shown as small circles bonded to the oxygen atoms of the hydroxyl groups. Rather than terminate the external plane of oxygen atoms of the octahedral sheet by hydroxyl groups, as in the 1:1 minerals, it is possible to add a second tetrahedral sheet, this one inverted with reference to the first tetrahedral sheet, making a 2:1 mineral structure. Here the "2" refers to the number of tetrahedral sheets and the "1" refers to the number of octahedral sheets. This type of mineral is shown in Figure 3.4, also in projection. Finally, there are layer structures which have octahedral (hydroxide) sheets inserted between 2:1 type layers. These are labeled 2:1:1 (see Figure 3.4c). Clearly, as the complexity of the layer increases, the thickness of the layer also increases. Since this thickness is easily measured by a simple X-ray diffraction experiment, the identity of the layer type is a trivial matter.

3.3.2 Octahedral site occupancy

While all the tetrahedral sites are occupied in these structures, albeit by several different cations, such is not necessarily the case for the octahedral sites. The octahedral sheet in all these structures is formed from two planes of closest packed (hexagonal) oxygen atoms and/or hydroxyl groups. These two planes are stacked in a closest packed arrangement forming an array of edge-sharing octahedra as shown in Figure 3.5. When all the octahedral sites are filled, as in Figure 3.5a, each oxygen is coordinated to 3 divalent cation (e.g., Mg^{2+}) or a total charge of +6. According to Pauling's electrostatic rule, this number of cations balances half of the -2 charge on the oxygen. The same situation occurs when the same oxygen is coordinated by 2 trivalent cations and an empty octahedral site (Figure 3.5b). The two extremes, all octahedral sites filled and 2/3 of the sites filled, occur so frequently that they are identified as "trioctahedral", literally 3/3 sites filled and "dioctahedral" for 2/3 sites filled. Thus, a 1:1 structure can exist as a dioctahedral or as a trioctahedral mineral. The same is true of the 2:1 structures but not of the 2:1:1 structures. While the problem of the charge neutralization of the oxygen atoms is solved equally well for the dioctahedral and trioctahedral cases, there are nonetheless important differences in the two structures. For example, Figure 3.5 shows that the octahedra in the trioctahedral sheet are regular and essentially equal in size (assuming that each is occupied by the same cation), such is not the case for the dioctahedral sheet. In the latter, the empty site creates an asymmetry of charge resulting in an increase in size of the empty site as the neighboring cations pull their coordinating oxygen closer. The asymmetry of cation charge also strongly influences the orientation of the various hydroxyls in the 1:1 and 2:1 structures, (as will be seen later) as well as the symmetry of the octahedral sheet.

3.3.3 Layer charge

The term "layer charge" refers to an excess of (usually) negative charge which resides on the layer as a whole. The origin of this charge is the substitution of cations in tetrahedral or octahedral (or both) sites which have a non-ideal charge. This can be understood in terms of ideal end-member structures which have no charge. These end-members are characterized by a 4+ cation in the tetrahedral sites and either a 2+ cation in a trioctahedral structure or a 3+ cation in a dioctahedral structure (Table 3.2). Note that the 2:1:1 structures can only occur where there is a negative layer charge, so this structure does not appear in the table.

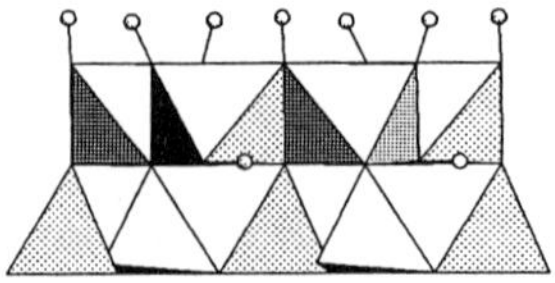

(a) 1:1 Layer

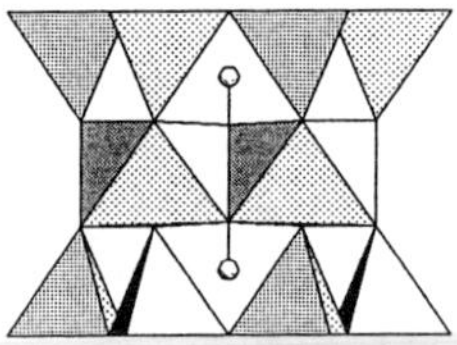

(b) 2:1 Layer

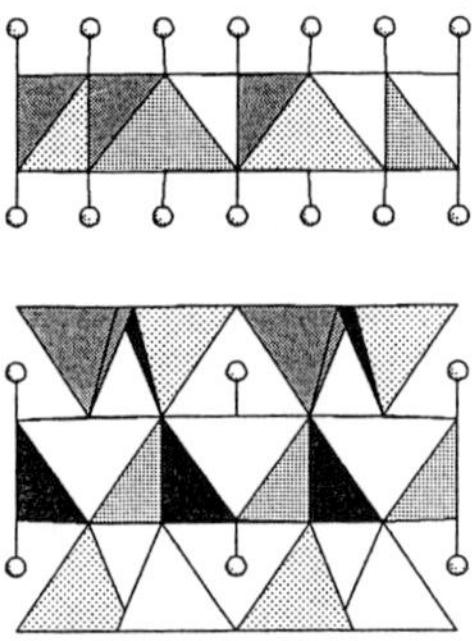

(c) 2:1:1 Layer

Figure 3.4: *Side views of the major types of layer silicate structures. The actual minerals shown are kaolinite, phlogopite 1M, and chlorite IIb-2. The small circles are the actual hydrogen atom positions.*

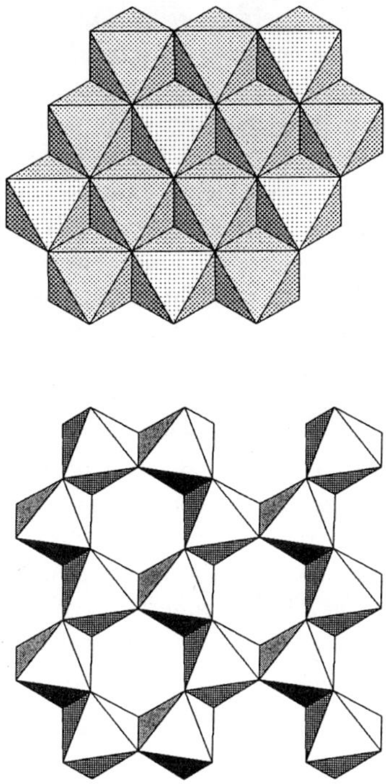

(a) Trioctahedral

(b) Dioctahedral

Figure 3.5: *Octahedral sheets for the trioctahedral (a) and dioctahedral cases (b). Note that the existence of the vacant octahedral site (b) creates distortions in the remaining octahedral sites and an enlargement of the octahedral site which is missing the cation.*

	Trioctahedral	Dioctahedral
1:1	$Mg_3Si_2O_5(OH)_4$ serpentine	$Al_2Si_2O_5(OH)_4$ kaolinite
2:1	$Mg_3Si_4O_{10}(OH)_2$ talc	$Al_2Si_4O_{10}(OH)_2$ pyrophyllite

Table 3.2: *Ideal end members for the 1:1 and 2:1 structure types for both dioctahedral and trioctahedral minerals.*

A negative layer charge can be created by the introduction of a cation with valence less than 4+ in the tetrahedral site or a cation with valence less than 3+ in a dioctahedral site or a cation with valence less than 2+ in a trioctahedral site. It is also clear that a positive layer charge can result from a 4+ cation in the dioctahedral case or a 3+ or 4+ cation in the trioctahedral structure. Examination of Table 3.1 shows that there are no 4+ cations which are normally comfortable in an octahedral site (note that Si^{4+} can occur in an octahedral environment at high pressure, the mineral stishovite, conditions where phyllosilicates are not often found). Therefore, the only common way in which a positive layer charge can be created is by the substitution of, e.g., Al^{3+} for Mg^{2+} in a trioctahedral mineral. There is a limit to this substitution because the oxygen bonded to three octahedral cations, which is completely charge balanced by three 2+ cations, will become charge unbalanced by the substitution of a 3+ cation. This imbalance ultimately affects the overall stability of the mineral. Thus, the predominance of observations shows that substitutions in both the tetrahedral and octahedral sites lead to a net negative layer charge. Further, the site of the charge may be tetrahedral or octahedral or, especially for the 2:1 clay minerals, both.

It is not surprising that the magnitude of the layer charge, and secondarily the site of the charge, can influence the chemical and physical properties of the phyllosilicate minerals, although it is frequently difficult to quantify this.

3.3.4 The interlayer

In order to have a stable mineral, there must be some degree of attraction between the layers making up the mineral. The discussion in this section is of a crystal-chemical nature; the interfacial aspects will be examined in a later chapter.

There is a fundamental difference between the 1:1 and the 2:1 layers, as has already been mentioned. The 2:1 layer is bounded on both sides by basal oxygen planes whereas the 1:1 layer has basal oxygens on one surface and hydroxyls on the other surface. The interlayer bonding for the 1:1 layer silicates, whether dioctahedral or trioctahedral, is via hydrogen bonds from one hydroxyl surface to the adjacent oxygen plane of the neighboring 1:1 layer. These are long hydrogen bonds, but there are many of them and thus their contribution to the interlayer bonding is strong. Any ionic substitution occurring in the 1:1 layer is usually such that overall electrostatic neutrality is maintained, i.e., the layer charge is always zero or very near to zero.

The 2:1 layer is more complex because it is possible to have a net layer charge. Since such a situation would be unstable because of the electrostatic repulsion between all the layers, the charge must be balanced by the presence of extra positive charge. There is no site in the layer for the placement of this charge (e.g., extra cations), but the positively charged entities easily can be situated between the layers. This creates an alternating arrangement of ...layer-interlayer-layer-interlayer... where the electrostatic charges also alternate, i.e., negative-positive-negative-positive. At first examination, the bonding between layers would seem to be a simple electrostatic attraction of oppositely charged entities. The real situation is more complex and will be discussed later.

There are basically four types of interlayer moiety commonly encountered in naturally occurring materials and as modified materials. These are: 1) a vacancy (for a zero-charge layer), 2) an inorganic cation (e.g., Na^{1+}), 3) an organic cation (e.g., a quaternary ammonium cation), and 4) an inorganic complex (e.g., a positively charged hydroxide sheet as in the 2:1:1 chlorite structure). The minimum layer charge is zero and the maximum is 2 (this is the charge per formula unit, $T_4O_{10}(OH)_2$; note that the negative sign is frequently omitted). Because the chemical and physical properties are strongly influenced by the layer charge, the magnitude of the charge is a major factor in the classification of the different types of 2:1 minerals. Table 3.3 shows this classification. Note that for the 1:1 structure type and for the zero charge 2:1 structures specific minerals have been listed because there are few minerals in these categories. The other names, e.g., smectite, are group names and are listed instead of specific minerals because of the great variety of individual minerals in each group. On rare occasions, interlayer water molecules are found in material with the kaolinite 1:1 layer; this variety is called halloysite and is not stable under ambient conditions, rapidly losing most of the interlayer water.

The interlayer bonding in the 1:1 structures is now well understood thanks to recent neutron diffraction structure studies which revealed the locations of the hydrogen atoms. Figure 3.6 shows two views of the full structure of kaolinite and Figure 3.7 shows similar views of lizardite (both as examples of dioctahedral and trioctahedral structures). Both structures are similar in that the hydroxide sheet contains hydroxyls (frequently termed the inner surface hydroxyls) which are oriented with respect to the 1:1 layer in different ways depending on the occupancy of the octahedral sheet. In lizardite the OHs are nearly perpendicular, while in kaolinite the hydroxyls are tilted slightly away from the perpendicular in spite of the marked asymmetry of the octahedral cations. The hydroxyl (termed the inner hydroxyl) which is situated in the oxygen plane shared by the octahedral and tetrahedral sheets is very sensitive to its environment. In lizardite, and, by analogy, all trioctahe-

Layer Type	Interlayer Occupancy (x = layer charge)	Dioctahedral	Trioctahedral
1:1	none (water) (x≈0)	kaolinite	serpentine
2:1	none (x≈0)	pyrophyllite	talc
	hydrated cations (x≈0.2-0.6)	smectites	
	hydrated cations (x≈0.6-0.9)	vermiculites	
	cations (x≈1)	(true) micas	
	cations (x≈2)	(brittle) micas	
2:1:1	hydroxide sheet variable x	chlorites 2:1:1	

Table 3.3: *The classification of the phyllosilicate minerals based on layer type, the nature of the octahedral sheet and the magnitude of the layer charge. For each group, a common mineral is listed. See sections 3.4.1 and 3.4.2 for more details on the phyllosilicate minerals.*

dral 1:1 structures, this OH is oriented perpendicular to the octahedral sheet and the hydrogen is in the center of the hexagonal arrangement of tetrahedra (Figure 3.7). In contrast, the inner hydroxyl of kaolinite is nearly in the plane of the shared oxygen plane. Finally, the distortion of the tetrahedra in kaolinite is easily seen in Figure 3.6. The distortion is due to an alternating rotation of the bases of neighboring tetrahedra, creating a ditrigonal symmetry rather than a purely hexagonal arrangement. The tetrahedral rotation (defined by the angle α) is a common feature of the tetrahedral sheets of all phyllosilicates. It originates in the dimensional misfit of octahedral sheets and tetrahedral sheets. In contrast, the amount of tetrahedral rotation in lizardite is considerably reduced compared to the dioctahedral mineral kaolinite.

Both the true and brittle micas are structurally similar, they differ only in having different layer charges. The interlayer cations necessary to balance the layer charge key into "dimples" in the surface of the 2:1 layers (Figures 3.8 and 3.9). These dimples result from the open structure of the basal

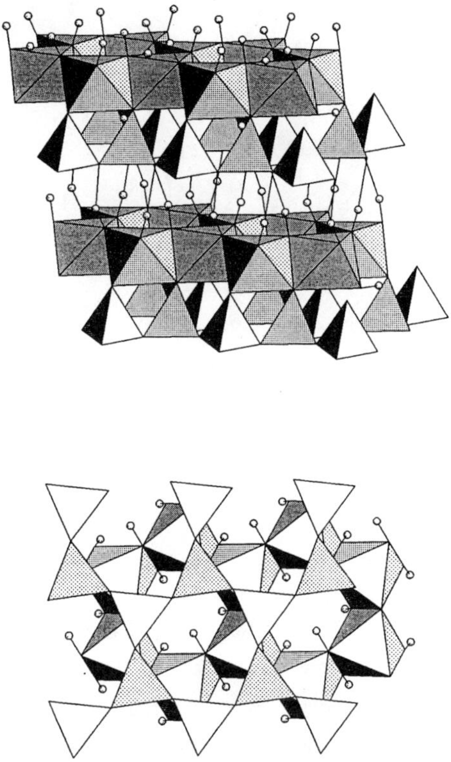

Figure 3.6: *A clinographic view of the (dioctahedral) kaolinite (upper) and a projection of the 1:1 layer (below).*

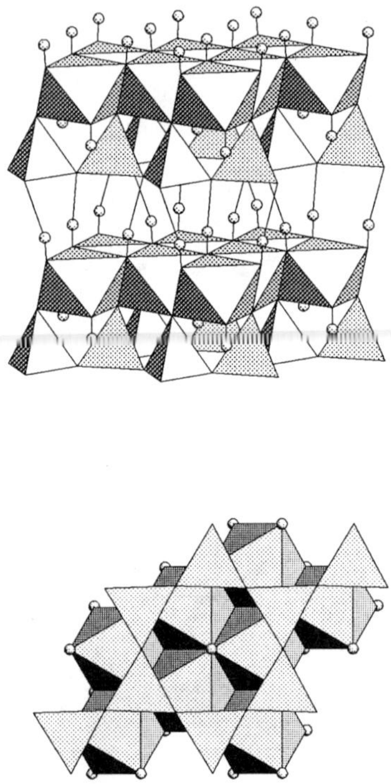

Figure 3.7: *A clinographic view of the (trioctahedral) lizardite structure (upper) and a projection of the 1:1 layer (below).*

oxygen atoms of the tetrahedral sheet. Thus, the position of the interlayer cation with respect to an adjacent 2:1 layer is fixed. This keying of the interlayer cation results in a high degree of registry of one layer with respect to the adjacent layers. There are, however, several different ways to stack 2:1 and 1:1 layers and this is normally indicated by a suffix attached to the mineral name. The details of these different stackings is not germane to the topic of this book because the external surface of a mica, for example, is the same and independent of the nature of the stacking of the layers. This arises because the origin of the different stackings lies at least 10 to 12 Å below the external surface and, thus, has a minuscule influence on the surface and interfacial interactions of the mineral. Nonetheless, it is worth explaining the manner in which the stacking is named. There are two contributors to this nomenclature; an integer followed by a capital letter, as in 1T. The integer refers to the thickness (i.e., the number of layers) of the repeating unit, in a crystallographic sense, and the letter refers to the symmetry of the stacking where T = trigonal, H = hexagonal, M= monoclinic, Or = orthorhombic and Tc = triclinic. Thus, the lizardite structure shown in Figure 3.6 is really lizardite 1T, that is a 1 layer repeat with trigonal symmetry. Occasionally, there is an integer subscript following the capital letter, as in $2M_1$, which indicates that there is more than one way to stack layers with the given symmetry and periodicity; these different ways are simply numbered.

There is only one type of hydroxyl in the 2:1 structures, the one situated in the plane of oxygen atoms shared by adjacent tetrahedral and octahedral sheets. This is structurally equivalent to the inner hydroxyl of the 1:1 structures. As is the case for the 1:1 trioctahedral minerals, the inner OH for the 2:1 trioctahedral minerals is perpendicular to the layer, assuming that all three cations bonded to the OH are identical (Figure 3.8). If these octahedral cations are not identical, especially if they have different charges, then the OH will deviate from the normal. In the case of 2:1 dioctahedral minerals, the OH lies just a few degrees away from the shared plane of oxygen atoms (Figure 3.9). This is a clear indication of the influence of the interlayer cation which is keyed into the ditrigonal hole and just a few Å away from the proton. Both being positively charged, there is a strong electrostatic repulsion which is relieved partially by the OH rotating away from the interlayer cation.

As the 2:1 layer charge decreases from that of true micas ($x \approx 1$) toward talc and pyrophyllite ($x \approx 0$), two major groups of phyllosilicates are encountered, vermiculites and smectites, in that order. It might be supposed that these two groups would have properties intermediate to the micas and talc/pyrophyllite. Such is not the case. Micas, on one hand, and talc/pyrophyllite, on the other hand, normally have well-ordered three-

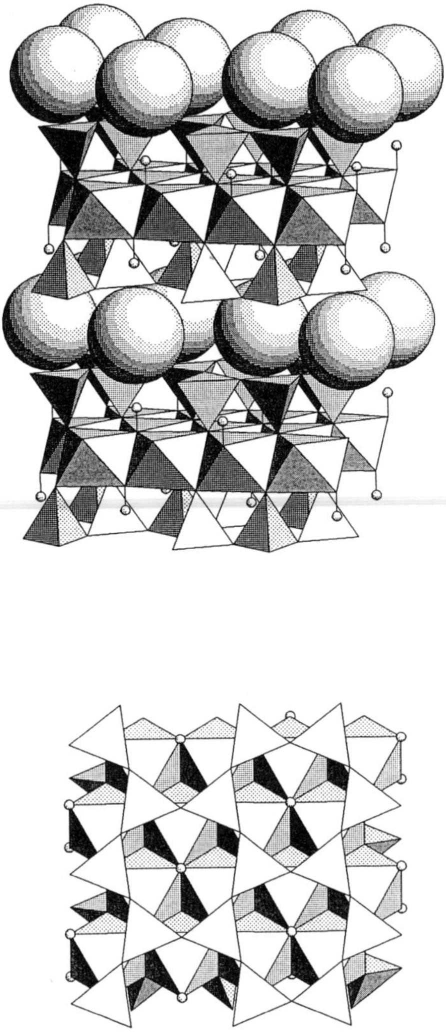

Figure 3.8: *A clinographic view of the (trioctahedral) phlogopite 1M structure (upper) showing the positions of the interlayer cations (large spheres) and a projection of a single 2:1 layer (lower, without the interlayer cations).*

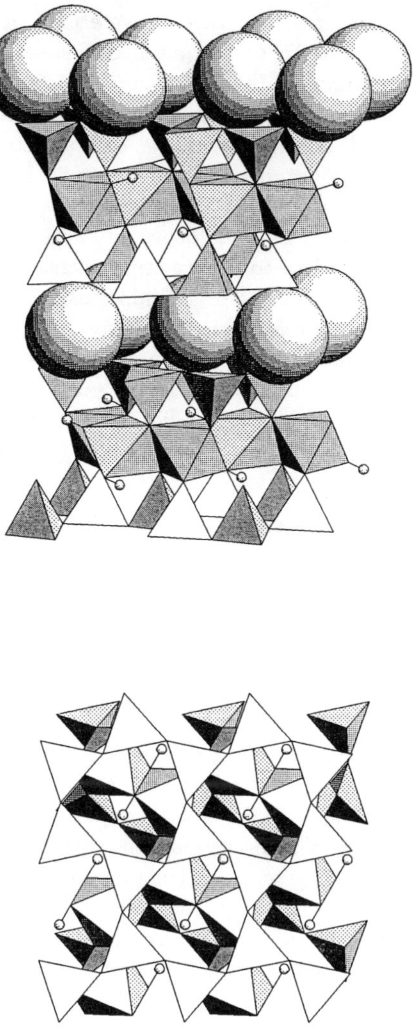

Figure 3.9: *A clinographic view of the (dioctahedral) muscovite $2M_1$ structure (upper) showing the positions of the interlayer cations (large spheres) and a projection of a single 2:1 layer (lower, without the interlayer cations).*

dimensional crystal structures. In contrast, smectites, especially those having the layer charge dominantly on the octahedral sheet, have a great deal of stacking disorder, and vermiculites exhibit a rich variety of stacking modes. Further, both groups normally have molecular water situated between the silicate layers, either coordinating the charge compensating cations or as free water. In addition, many organic molecules can be introduced very easily into the interlayer space, displacing the interlayer water. Such behavior is not known for either the micas or talc/pyrophyllite. Smectites differ from vermiculite by their ability to incorporate large quantities of interlayer water, again, especially for octahedrally charged minerals.

Because of the stacking disorder in smectites, we know relatively little about the structure of the interlayer hydrated cations and how these relate to the adjacent silicate layers. We do have some information about the structure of vermiculites, and these studies are frequently taken as models for the interlayer (and surface layers) of smectites (see section 3.4.2).

3.3.5 Chemical variations

A word must be said about the chemical composition of a complex mineral such as a smectite and how it relates to the structural formula. The two are very different; the chemical analysis, whether done by traditional wet chemical methods or by X-ray fluorescence or by microprobe or some other technique, provides a list of the weight percentages of the metal oxides and water which are present in the sample. Oxygen normally is not analyzed. For pure materials, the conversion of the oxide analysis to a structural formula is a trivial operation. But clay minerals, and even many micas, are not pure materials; the problem lies in how to apportion those elements (e.g., aluminum) which can occur in more than one structural site. In addition, the analysis does not indicate whether the mineral is dioctahedral or trioctahedral or a mixture of the two. In this sense, a vacancy is treated as an octahedral cation with zero valence. As an example, Table 3.4 presents the oxide chemical analyses of two minerals, one, a magnesium-rich olivine (sample 1 from Deer *et al.*, (1972), page 10), is straight forward while the second, illite, a 2:1 mica-like mineral, is more complex (sample 1, from Weaver and Pollard, (1973). The columns in the table are largely self explanatory. The weight % is given by the chemical analysis, the atomic weights are standard values, and the atomic proportion is the ratio of the weight % to the molecular weight of the oxide. The atomic proportions are recast in terms of the cation atomic proportion (column 5) and the oxygen atomic proportion (column 6). The oxygen column is summed and a factor (1.409 in this case) calculated as the ratio of the ideal oxygen content of the formula (4 in this case) divided by the total oxygen atomic proportion (2.839). Actually, what is of interest is the number of negative charges to be balanced by the total of the cation content. For olivine, there are 8 negative charges to be balanced, and that is the same as 4 oxygen atoms. This factor is used to scale the cation and oxygen columns to give the last two columns in the table. The values for the recalculated cation content (column 7) are the cation contents of the structural formula. There may seem to be a deficiency of Si but 0.978 is probably not significantly different from 1. Similarly, the sum of the ferrous iron and magnesium, 2.044, is not significantly different from 2. One could write the structural formula with these determined values, as is done in the table, or simply renormalize to the expected values, 1 for Si and 2 for Mg + Fe.

The conversion of the illite (from Fithian, Illinois) chemical analysis into a structural formula is not as straightforward as was the olivine example. The complications derive from the number of different cations reported, many in relatively small but not negligible amounts, the possibility that the material is not a purely dioctahedral mineral, and the presence of two kinds of water, H_2O^+ and H_2O^-. The superscript "-" indicates that the water was lost from the mineral sample at relatively low temperatures, e.g., 140°C. This kind of water is, in phyllosilicate minerals, either surface adsorbed water or interlayer water which is coordinating charge compensating cations. With proper sample preparation, surface adsorbed water is usually not present. Water which is lost at very high temperatures, e.g., 1000°C, is reported as H_2O+. This is not really water but hydroxyl groups which are part of the structure. In the case of illite, the number of negative charges to be balanced is 22, since the ideal formula is $M_2T_4O_{10}(OH)$. This negative charge can also be treated as 11 oxygen atoms, for simplicity.

Having scaled the cations (column 7), it is clear that there are not enough silicon ions to fill the 4 tetrahedral sites, so the missing material is drawn from the other cations in analysis. Here a bit of judgment is required, and the normal assumption is that aluminum will be preferred over other cations such as ferric iron. In the present analysis, there is more than enough aluminum to bring the cation total to 4 (3.43 Si + 0.57 Al). The remaining aluminum occupies the octahedral sites along with titanium, ferrous and ferric iron and magnesium. This gives a total octahedral occupancy of 2.12 cations, indicating a small amount of trioctahedral character. There remain some alkali and alkaline earth oxides which do not fit in the structure of the layer. These go into the interlayer along with the free water. Finally, the high temperature water is reported as hydroxyl, of which there is slightly less

Olivine Chemical Analysis

Oxide	Weight %	Molec. Weight	Molec. Prop.	Cations (relative)	Oxygen (relative)	Recalc. Cations	Recalc. Oxygen
SiO_2	41.72	60.08	0.694	0.694	1.389	0.978	1.957
FeO	1.11	71.85	0.015	0.015	0.015	0.022	0.022
MgO	57.83	40.30	1.435	1.435	1.435	2.022	2.022
Totals	100.66				2.839	3.022	4.000
Multiplicative factor = 1.409 (basis of 4 oxygens)							
Structural formula = $(Mg_{2.02}Fe_{0.02})Si_{0.98}O_4$							

Illite Chemical Analysis

Oxide	Weight %	Molec. Weight	Molec. Prop.	Cations (relative)	Oxygen (relative)	Recalc. Cations	Recalc. Oxygen
SiO_2	51.22	60.08	0.852	0.852	1.705	3.433	6.867
TiO_2	0.53	79.90	0.007	0.007	0.013	0.027	0.053
Al_2O_3	25.91	101.96	0.254	0.508	0.762	2.047	3.070
Fe_2O_3	4.59	159.69	0.029	0.057	0.086	0.232	0.347
FeO	1.70	71.85	0.024	0.024	0.024	0.095	0.095
MgO	2.84	40.30	0.070	0.070	0.070	0.284	0.284
CaO	0.16	56.08	0.003	0.003	0.003	0.011	0.011
Na_2O	0.17	61.98	0.003	0.005	0.003	0.022	0.011
K_2O	6.09	94.20	0.065	0.129	0.065	0.521	0.260
H_2O^+	7.14	18.02	0.396	0.793	-	-	1.596
H_2O^-	1.45	18.02	0.080	0.161	-	-	0.324
Totals	101.80				2.731	6.672	11.000
Multiplicative factor = 4.028 (basis of 11 oxygens)							
Structural formula = $[Ca_{0.01}Na_{0.02}K_{0.52} \cdot 0.32H_2O]$ $[Al_{1.48}Fe^{3+}_{0.23}Fe^{2+}_{0.10}Mg_{0.28}Ti_{0.03}]$ $[Si_{3.43}Al_{0.57}]O_{10}(OH)_{1.6}$							

Table 3.4: *Derivation of a structural formula from oxide weight percents. See the text for details of the cation assignment to the different structural units.*

than the ideal amount.

This process may seem simple, but frequently there is a degree of arbitrariness in assigning cations to different structural sites which is a matter of judgment. For example, there is no unambiguous way to determine how much of the total aluminum is really in the octahedral and tetrahedral sites.

Having decided on a structural formula, the layer charge can be determined. For the illite in this example, the contribution from the tetrahedral sheet is -0.57 and the octahedral sheet contributes +0.00. The charge on the interlayer cations is +0.57, which just balances the layer charge, giving one some confidence in the correctness of the structural formula. Frequently, the numbers do not work out quite this well.

3.4 Phyllosilicate Minerals

Nomenclature is often a problem particularly when one looks in the older literature. As our knowledge of the structure and chemistry of clay minerals has improved since the first descriptions of the 19th Century, names have been discarded or it was found that what was once treated as a single mineral was actually a group of related minerals.

As was shown in Table 3.3, the primary characteristic of the layer silicates is the architecture of the silicate layer itself. This property distinguishes the 1:1, 2:1, 2:1:1 groups. Because the nature of the layer is so directly reflected in the thickness of the layer, these groups are also referred to as the 7 Å, 10 Å and 14 Å groups, where the thickness represents the stacking of identical layers with no water or other molecular species occupying the interlayer region.

Specific minerals in each group are distinguished by a combination of chemical composition, whether the octahedral sheet is dioctahedral or trioctahedral, the existence or absence of interlayer molecular species (principally water), and by the degree of disorder in the layer stacking. A full discussion of these factors is not necessary here; the interested reader should look at Moore and Reynolds (1997). The following will discuss only the most common clay minerals and their properties, especially those which may be related to surface properties.

3.4.1 1:1 minerals

The minerals belonging to this group have a characteristic layer thickness of 7 Å or slightly larger. The major differences between the minerals in this group are the manner in which the octahedral sites are filled, the type of cations filling the tetrahedral and octahedral sites, and, finally, the manner in which the layers are stacked. In fact, the layer stacking is quite complex, but there presently seems to be little or no relation between the different stackings and the surface properties of the minerals. There is much less chemical substitution in the 1:1 minerals than appears in the 2:1 minerals, so the variety of physical and chemical properties of the former minerals are more restricted.

Trioctahedral

These minerals are commonly referred to by the group name serpentine with the ideal chemistry $(Fe,Mg)_3Si_2O_5(OH)_4$. There can be substitution of Al^{3+} for both the tetrahedral and octahedral cations leading to a positive charge on the tetrahedra and an equivalent compensating negative charge on the octahedra as in the mineral amesite. For all these materials, the total layer charge is 0, there are no interlayer cations and the layers are held together by hydrogen bonding. There has been no report of the intercalation of water or organic materials between the layers of the serpentine minerals. The interlayer bonding of these minerals is apparently very strong.

Under special geological conditions, mild metamorphism of rocks rich in serpentine, the normally flat 1:1 layers will roll into tubes. The classical explanation for this is that there is a geometrical misfit between the lateral dimensions of the octahedral and tetrahedral sheets, the octahedral being slightly larger. This misfit leads to a curling and, when continued to the extreme, the result is a long rolled sheet. This morphology leads to the formation of thin and long (asbestiform) fibers. In such a form the mineral is commonly called chrysotile (the so-called white asbestos). There is still controversy concerning the ability of chrysotile to lead to cancer when small fibers are introduced to the human lung. This question will be treated in Section 10.5.

Details of the structure and chemistry of these minerals are discussed by Wicks and O'Hanley (1982).

Dioctahedral

A commonly used group term for all these materials is kaolin. Kaolin minerals have an ideal formula of $Al_2Si_2O_5(OH)_4$. There are three widely recognized polytypes: kaolinite (the most common), dickite, and nacrite (the rarest). These three minerals differ in having a different stacking sequence of the 1:1

layers. Kaolinite, and likely the other minerals as well, exhibits a wide range of stacking disorder. Again, there is little evidence now that the nature of the stacking leads to important differences in the surface properties of these minerals.

Unlike the serpentine minerals, a large variety of molecular species can be introduced between the 1:1 layers of kaolinite. The organic molecules tend to be small and highly polar, e.g., formamide and dimethyl sulfoxide. These intercalates may be highly ordered materials and offer a useful model for studying the interaction of organic molecules with oxide surfaces. In addition to the molecular species, certain inorganic salts also can be introduced between the 1:1 layers (sometimes called intersalates). None of these intercalates or intersalates has been recognized in natural mineral deposits.

Water frequently forms an intercalate with kaolinite, in the form of the mineral halloysite which has the formula $Al_2Si_2O_5(OH)_2 \cdot 2H_2O$. The addition of the two molecules of water increases the thickness of the layer from 7 Å to 10 Å. Kaolinite itself can be hydrated by a relatively simple chemical treatment to produce both a dihydrate, similar to halloysite, and a monohydrate with a thickness of about 8.6 Å. Kaolin minerals do not form fibers as readily as do the serpentines. These minerals are described in more detail by Giese (1988).

3.4.2 2:1 minerals

The identity of the 2:1 minerals is determined by the magnitude of the layer charge, resulting from the substitution of cations of different formal charge in the tetrahedral and octahedral sites. In general, the higher the layer charge, the stronger the interlayer bonding, the greater the difficulty of introducing molecular materials between the layers and the less likely will be the exchangeability of the interlayer cations. The following sections discuss the different mineral groups in order of decreasing layer charge.

Micas

Micas commonly occur in large tabular crystals (sometimes measuring several feet in diameter) exhibiting a characteristic excellent cleavage parallel to the tabular habit. In crystals low in defects and imperfections, extremely thin sheets of mica can readily be split using the edge of a sharp blade.

The primary source of the layer charge in the micas is the substitution of di- and trivalent cations (largely Al^{3+} for silicon in the tetrahedral sites). The maximum substitution of aluminum for silicon occurs when half the tetrahedral sites are occupied by aluminum and half by silicon. This yields a layer charge of +2 for the $T_4O_{10}(OH)_2$ unit. Such minerals are termed brittle micas because the cleavage sheets break when flexed. The +2 layer charge is balanced by divalent interlayer cations. The most common brittle mica is margarite, $CaAl_2(Al_2Si_2)(OH)_2$.

If the substitution of aluminum for silicon has the ratio 1/3, the mineral is a mica with a layer charge of +1. These are geologically very common materials typically forming under high temperature and pressures in igneous or metamorphic rocks. The micas and their degradation products commonly occur in sediments and sedimentary rocks. The major dioctahedral micas are muscovite, $KAl_2(AlSi_3)O_{10}(OH)_2$, and paragonite, the sodium equivalent, $NaAl_2(AlSi_3)O_{10}(OH)_2$. These have the full mica layer charge on the tetrahedral sites. An intermediate situation, where ideally half the charge on both the tetrahedral and octahedral sites is the mineral phengite, $K(Mg_{0.5}Al_{1.5})(Al_{0.5}Si_{3}.5)O_{10}(OH)_2$. Trioctahedral micas have +2 cations in the octahedral sites, principally magnesium and ferrous iron. For a tetrahedral charge, the iron rich mica is annite, $KFe_3(AlSi_3)O_{10}(OH)_2$, the magnesium rich variety is phlogopite and intermediate compositions are biotite, $K(Mg,Fe)_3(AlSi_3)O_{10}(OH)_2$. If the layer charge lies on the octahedral sites, the dioctahedral mineral is celadonite, $K(MgAl)(Si_4O_{10}(OH)_2$. For the trioctahedral case the mineral is taenolite $K(Mg_2Li)Si_4O_{10}(OH)_2$ (not a common mineral but an important end member).

As mentioned earlier, illite is a mica-like structure with a slightly reduced layer charge. The difficulties in determining exactly what illite is, in terms of chemistry and structure, are related to the existence of (hydrated) smectite layers randomly interspersed in the illite layers. It is not easy to identify the existence of small quantities of smectite layers in an illite sample. Illite presently is defined as "a non-expanding, dioctahedral, aluminous, potassium mica-like mineral" of small grain size (Srodon and Eberl, 1984). The exact value of the layer charge for illite has been a matter of contention. A value of -0.75 has been accepted, although there is evidence that there may be two types of illite layer, one having a charge of -0.55 and the other a higher charge of -1 (a mica charge). When present in roughly equal quantities in an illite sample, the overall charge would be close to -0.75.

Vermiculites

This is a very complex group of materials. They frequently occur in macroscopic crystals, although not nearly as large as the micas from which they are generally derived by chemical modification. These minerals are normally trioctahedral (the so-called true vermiculites) and have an ideal chemical formula of $Mg_{0.6}$ $n(H_2O)$ $(Mg,Al,Fe^{2+})_3(Si,Al)_4(OH)_2$ where the items in the square brackets represent the interlayer cations (principally Mg) and interlayer water (n depends on the relative humidity and the abundance of the interlayer cation). The layer charge varies between 0.6 and 0.9 (the formula corresponds to a charge of 0.9) and arises from substitutions in both the tetrahedral and octahedral sites. There is a great deal of variety in the stacking sequences of these minerals (de la Calle and Suquet, 1988). The fact that vermiculites frequently occur in macroscopic crystals provides the opportunity to study these minerals and particularly the placement of interlayer cations and water or organic molecules. Much of what we know about the physical interactions of molecular species at the mineral-liquid interface has been derived from studies of vermiculites. To the extent that this interface is the same as that for smectites and other silicate minerals, the results are useful. These results will be discussed in a later section.

Smectites

The smectite minerals are distinguished according to the nature of the octahedral sheet (dioctahedral versus trioctahedral), by the chemistry of the layer and by the site of the charge (tetrahedral versus octahedral). In reality, the smectite minerals are a very complex group, frequently having both octahedral and tetrahedral substitutions each contributing to the overall layer charge. Thus, the mineral names represent ideal end members. The following formulae are based on a layer of 0.33, with the value varying between approximately 0.2 and 0.6 and sodium is indicated as the interlayer cation.

There are three important dioctahedral smectites; two are aluminous and the third is iron rich. With a predominantly octahedral charge, the mineral is montmorillonite, $Na_{0.33}$ nH_2O $(Al_{1.67}Mg_{0.33})Si_4O_{10}(OH)_2$ where, as before, the n indicates a variable amount of interlayer water coordinating (or not) the interlayer cation. The interlayer cation need not be sodium, but this is the common occurrence. It should be noted that in the older literature the term "montmorillonite" was frequently used as a group name for any swelling 2:1 clay mineral as well as the name of a specific mineral, clearly not a good situation. Presently smectite is the group name and montmorillonite is restricted to a mineral name belonging to that group. If the charge is predominantly tetrahedral and aluminous the mineral is beidellite with an ideal composition of $Na_{0.33}$ nH_2O $Al_2(Al_{0.33}Si_{3.67})O_{10}(OH)_2$. Finally, if ferric iron substitutes for aluminum in the octahedral sites and the charge is tetrahedral, one has nontronite, $Na_{0.33}$ nH_2O $Fe_2^{3+}(Al_{0.33}Si_{3.67})O_{10}(OH)_2$.

The trioctahedral equivalent of montmorillonite is the mineral hectorite, ideally $Na_{0.33}$ nH_2O $Mg_3(Al_{0.33}Si_{3.67})O_{10}(OH)_2$. It should be noted that in contrast to montmorillonite, hectorites have lithium (1+) in some octahedral sites (not shown in the above formula) adding to the total layer charge. Saponite is the trioctahedral equivalent of beidellite but differs in that some aluminum substitutes for magnesium in the octahedral sites generating a positive contribution to the layer charge which reduces the negative contribution from the tetrahedral sites. The ideal formula without the aluminum substitution is $Na_{0.33}$ nH_2O $Mg_3(Al_{0.33}Si_{3.67})O_{10}(OH)_2$.

In reality, there is often a range of compositions, for example between montmorillonite and beidellite. Some of these relations can be seen in Figure 3.10.

The interactions between adjacent smectite layers are not very strong and the interlayer material, hydrated cations, water, organics, are disordered so that there is little coherence from one layer to the next. This situation is described as turbostratic. As a consequence, it is normally not possible to speak of a crystal of a smectite. There are some exceptions, saponite being one, where there is a greater degree of stacking regularity. Given this degree of disorder, it is difficult if not impossible to study the physical structure of the water (or organics) in contact with the silicate layers. Thus, our knowledge of the interface for smectites is not great, and one relies on the vermiculite hydrates and organic intercalates for structural information (see section 3.4.2).

Talc and pyrophyllite

There are only two 2:1 minerals with zero layer charge. The trioctahedral mineral is talc, $Mg_3Si_4O_{10}(OH)_2$ and the dioctahedral mineral is pyrophyllite, $Al_2Si_4O_{10}(OH)_2$. These minerals are normally found in metamorphic environments rather than the sedimentary environment typical of clay minerals. Normally, these are reasonably pure compounds with little substitution by other elements. Thus, the layer charge is zero or very small and the layers are held together by a combination of Lifshitz-van der Waals forces along with a net ionic attraction, in spite of the insignificant layer charge (Giese *et al.*, 1991). These are normally thought of as weak forces, yet there is no report in the literature describing the intercalation of water or any sort of organic molecular species between the layers. There is no intercalation of water in these two minerals because of the hydrophobicity of the 2:1 layers. The contact angel of water on these minerals, θ_{water}, is too great for water to penetrate between the layers: they are *waterproof*! On the other hand, the

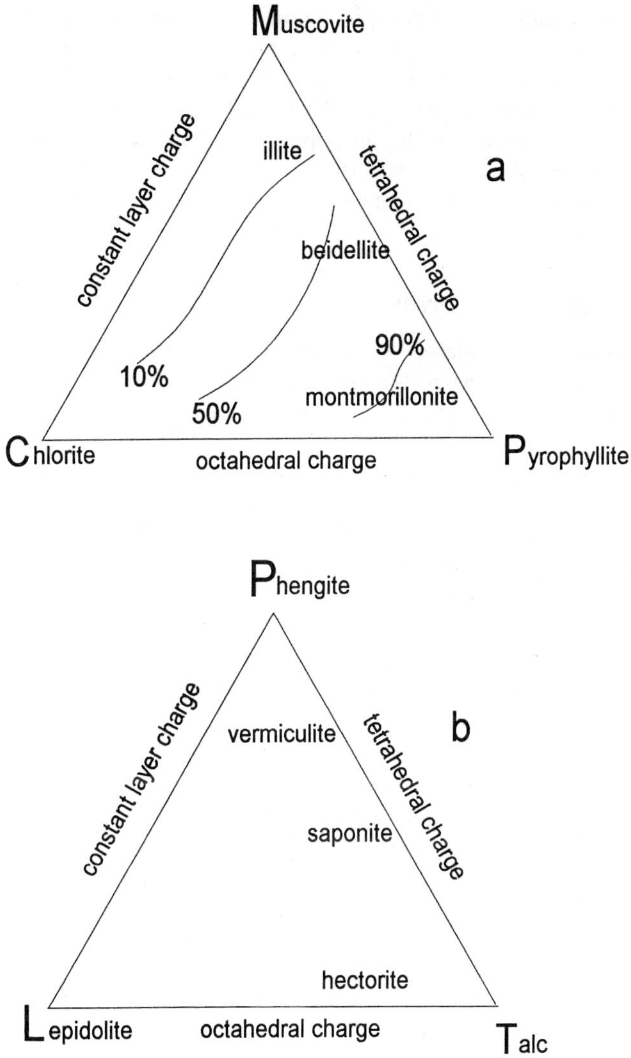

Figure 3.10: *The stability fields for dioctahedral (a) and trioctahedral (b) 2:1 structures as a function of layer charge and the site of the layer charge. The figures are for aluminous structures (a) and magnesian structures (b).*

lack of intercalation by hydrophobic organic liquids, e.g., pentane, is likely due to the relatively large size of these molecules and the energy required to separate the individual layers of talc or pyrophyllite.

These two minerals are frequently taken as the model for all 2:1 minerals, the others being generated by the appropriate ionic substitution in octahedral and tetrahedral sites along with the addition of exchangeable cations between the layers as necessary to balance the layer charge created by these structural substitutions.

3.4.3 2:1:1 minerals

These minerals are related to the 2:1 structures by replacing the exchangeable interlayer cations of the latter by a layer or partial layer of cations coordinated by hydroxyls, in a manner reminiscent of the octahedral sheet of the 2:1 layer itself. Thus, the (negative) layer charge of the 2:1 layer is compensated by a (positive) interlayer hydroxide layer. The presence of a hydroxide layer between the 2:1 layers creates a structure with a characteristic thickness of 14 Å. The group name is chlorite and most of these minerals are trioctahedral in both the 2:1 part of the structure and in the interlayer hydroxide layer. The dominant octahedral cations are generally in the magnesium and iron series. It is easily appreciated that one must take into account the dioctahedral/trioctahedral nature of the octahedral part of *both* the 2:1 layer and the interlayer hydroxide layer. It is not a simple matter to translate a chemical analysis into an accurate structural formula because many of the same cations appear in both layers and the partitioning of e.g., Mg between the 2:1 layer and the hydroxide layer is problematic. As an example, the magnesium trioctahedral chlorite is the mineral clinochlore, $(Mg,Al)_3(OH)_6$ $(Mg,Al)_3(Si_3Al)O_{10}(OH)_2$. This formula is sometimes represented as $(Mg_5Al)(Si_3Al)O_{10}(OH)_8$, when the interlayer hydroxide is included with the octahedral sheet of the 2:1 structure. In either representation, it is seen that the substitution of a cation with charge $> 2^+$ (e.g., aluminum) for the divalent octahedral cations in the dioctahedral 2:1 sheet as well as the interlayer hydroxide layer create a positive charge. In the 2:1 layer, the aluminum for silicon substitution, in this example, creates a negative charge which dominates over the positive contribution from the 2:1 octahedral sites. Details of these minerals are discussed by Bailey (1988a).

3.5 Interlayer Water

The foregoing discussion indicated that several phyllosilicate minerals, either naturally or as the result of chemical treatment, have molecular species inserted between the silicate layers. Water is the most common interlayer species in nature, and water is normally found in smectites, vermiculites and hydrated halloysites. The quantity of interlayer water is a function of relative humidity and the type of interlayer cation, in the case of smectites and vermiculites.

There is a great interest in the nature of the interface between water and silicate minerals (see for example Davis and Hayes (1986)). Much of the chemical activity in soils, sediments and porous rocks occurs at such an interface. Experimentally, it is very difficult to examine this interface because it is such a small part of the liquid-solid system. Hydrated smectites and vermiculites have water between all of the silicate layers and therefore the percentage of the sample which is interface is enormously larger than the interface between, for example, a grain of quartz in contact with liquid water. Another way to look at this is that the surface are of a quartz sand is probably much less than 1 m^2/gram while a typical smectite has a surface area of as much as 800 m^2/gram. For these, and other reasons, intercalated clays have been extensively studied.

3.5.1 Structure of interlayer water

As mentioned earlier, the stacking of smectite 2:1 layers is highly disordered, making the study of intercalated water and organic species difficult but not impossible. Vermiculites frequently exhibit a much higher degree of stacking order which, combined with the larger crystal size, allows traditional single crystal structure studies. Even with a high degree of layer stacking order, there is frequently disorder, both in terms of the position of the molecules and their orientation, so that a structure determination derived from diffraction data shows only an average structure. Vermiculites, in contrast, have a higher charge and the charge is frequently situated in the tetrahedral sites. This combination seems to lead to a greater degree of coherence in the stacking of the layers, hence a greater degree of three dimensional order allowing crystal structure determinations. Even though vermiculites are much better ordered than are smectites, there still is a large degree of randomness, particularly in the placement of molecular species between the layers.

Slade *et al.* (1985) have refined the structures of the Na and Ca forms of the Llano vermiculite, a mineral which occurs in large crystals having a relatively high degree of structural order. This is a particularly interesting study because it allows a direct comparison of the structure of vermiculite in either the monovalent or the divalent form. The Llano vermiculite has the majority of the layer charge situated in the tetrahedral sites. One might suppose that the close proximity of the layer charge to the interlayer region would control the placement of the interlayer cations, such that the preferred interlayer cation site would be directly above the tetrahedral sites. On the other hand, the ditrigonal hole of the silicate layer provides a geometrically advantageous position for the interlayer cation.

The structure refinement showed that the placement of the Na interlayer cation is different from that of the Ca. There are three possible sites for the interlayer cations: two of these are situated over three oxygen atoms which form the base of a tetrahedron (there are two different tetrahedra in the Llano vermiculite) and the third site is above the ditrigonal hole.

The placement of the interlayer cations appears to be determined by several factors. The first and most obvious is the attraction between the charge deficit situated in the tetrahedral sites and the nearby interlayer cations. Secondly, the position of the interlayer water molecules restricts the possible positioning of an interlayer cation. The interlayer water molecules are involved in a complex arrangement of hydrogen bonds, partly from water molecule to water molecule and also from water molecule to oxygen atoms of the silicate surface. The strength of these bonds produces a relatively rigid arrangement of water molecules in the interlayer region with a basically closest packed arrangement (Farmer and Russell, 1971).

Sodium has a relatively low polarizing capacity because of its small charge (1+) and relatively large size (1.02 Å for octahedral coordination) so it has little ability to distort the water molecule positions. This results in the sodium being octahedrally coordinated by water molecules (Farmer and Russell, 1971; Slade *et al.*, 1985). There is a clear indication that the sodium is situated as close as possible to the source of the layer charge. Also, the sodium is placed equidistant from the two adjacent layers, i.e., in the middle of the interlayer space (Figure 3.11). The structure refinement also shows that the site above the ditrigonal holes is vacant (Figure 3.11).

Calcium has approximately the same ionic radius as sodium (1.00 Å for octahedral coordination) so the calcium is better able to rearrange the water molecules in the interlayer volume than is sodium. The structure of the Ca-vermiculite shows that the three possible cation sites are occupied by Ca^{2+} ions. One site, termed m_1, has the calcium octahedrally coordinated by water in a rather regular fashion (Figure 3.12). The second type of Ca^{2+} situated above the tetrahedral sites, m_2, is similarly coordinated by six water

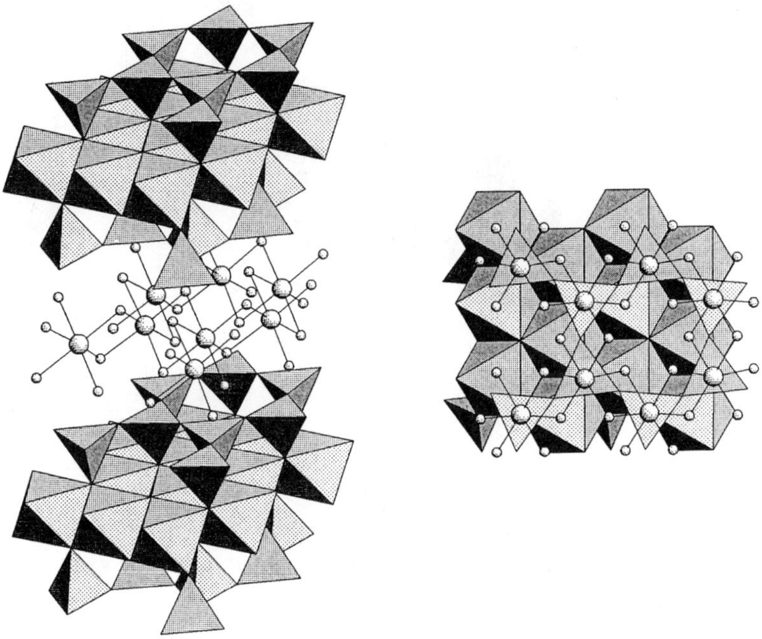

Figure 3.11: *The coordination of water molecules about the Na atom situated above the basal tetrahedral oxygen atoms. The left view shows an oblique view of the interlayer water structure and the right is a projection onto the layer of the same Na-water coordination arrangement showing the placement of the Na atoms above the tetrahedra of the adjacent silicate layer. In these illustrations, only the* Na^+ *ions and the coordinating water oxygen atoms are shown; hydrogen atoms have been omitted.*

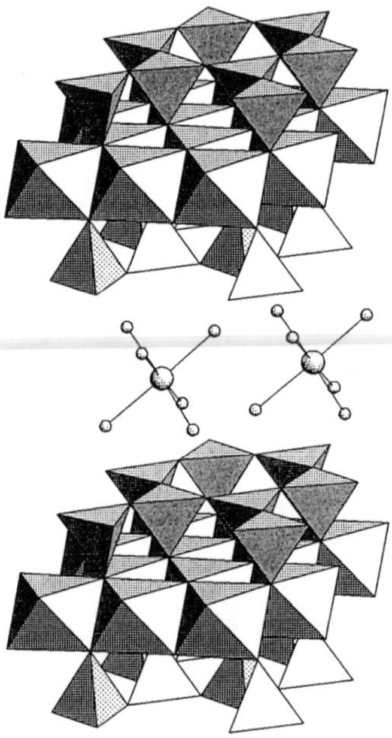

Figure 3.12: *The coordination of water molecules about the m_1 Ca atom situated above the basal tetrahedral oxygen atoms. The coordination is a reasonably regular octahedron.*

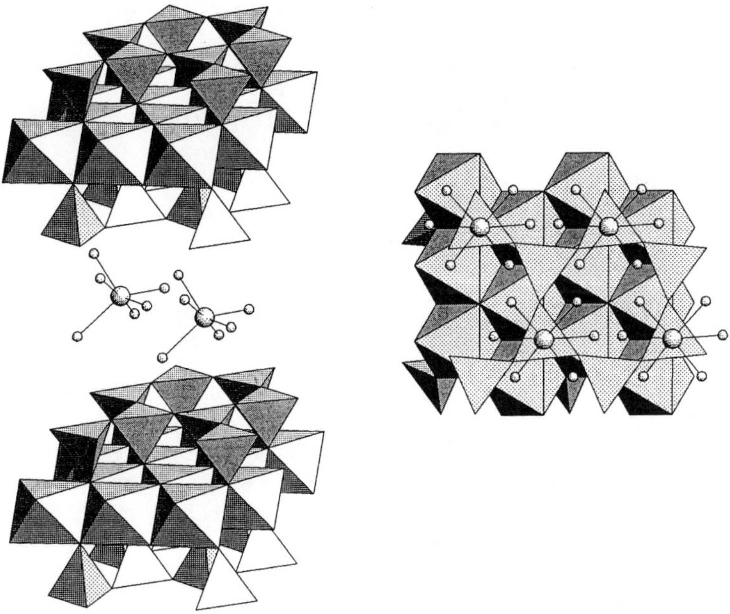

Figure 3.13: *The coordination of water molecules about Ca^{2+} the m_2 atoms situated above the basal tetrahedral oxygen atoms. The coordination about the calcium is a dramatically distorted octahedron.*

molecules but the coordination is strongly distorted from the ideal octahedron (Figure 3.13). With the exception of the distortion, the m_1 and m_2 calcium ions and their coordinating water molecules play the same role as do the sodium atoms. What sets the calcium vermiculite structure apart from the sodium vermiculite is the third cation site, m_3, which lies above the ditrigonal hole (Figure 3.14). This site has a greater volume available because of the proximity of the ditrigonal hole. The coordination number about this calcium ion is eight, and two of the water molecules are keyed into the two adjacent ditrigonal holes which face each other across the interlayer volume.

3.6 Intercalated Organic Molecules

The surfaces of clay minerals present a number of potential sites at which organic molecules could attach themselves. These sites include the exchangeable cations, water molecules coordinating the exchangeable cations as well as water molecules outside of the coordination sphere of the exchangeable cations, the oxygen atoms of the 001 surface of the silicate layer (here we must distinguish between the oxygen atoms that are adjacent to the substituting cation which is creating a charge imbalance as well as those which are not directly involved in the layer charge), hydrogen atoms which are part of surface hydroxyl groups (in the 1:1 and 2:1:1 structures), oxygen atoms which are at the edge of layers where bonds have been broken (including hydroxyl and water molecules replacing missing oxygen atoms at the broken edge), and any preexisting organic material adsorbed on external particle surfaces and intercalating molecules. Clearly there is a wide variety of interaction possible between the heterogeneous clay surface and the different functionalities of organic materials.

The only detailed information available on the structures of organo-clays is from studies of vermiculite and kaolinite complexes. Unfortunately, smectites do not form three-dimensionally ordered structures. For the organo-smectites, it is possible to roughly determine the orientation of the molecules in the interlayer space based on the dimensions of the molecules, assumptions as to how the molecules are attached to the clay surface and measured increases in the interlayer spacing compared to the untreated clay mineral. In certain cases it is possible to calculate the X-ray diffraction intensities for model structures and compare these with measured values. For example, Reynolds (1965) showed that the ethylene glycol molecule is oriented with the symmetry plane of the molecule oriented normal to the clay layer and coordinating the interlayer Ca^{2+} ions in a structure with only one-dimensional

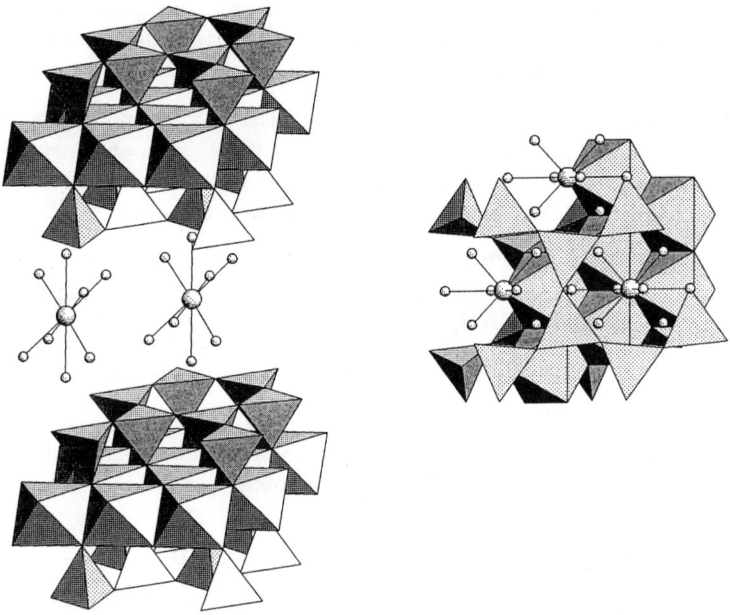

Figure 3.14: *The coordination of water molecules about the m_3 Ca^{2+} atoms situated above the ditrigonal hole in the silicate layer. The coordination is larger than for the m_1 and m_2 calcium atoms and very distorted.*

ordering (i.e., along the stacking direction of the layers).

One can categorize the types of organic-clay interactions based on the bonding mechanisms between the organic and the clay surface/interlayer (inorganic) cations. For example, MacEwan and Wilson (1980) list the following common types of interactions (covalent bonding between the organic and the clay surface is not included).

- Cationic Bonding: These involve organic cations such as the alkyl ammonium cations or amines and carbonyl groups which have become protonated, depending on the pH.

- Ion-dipole and Coordination Bonding: This is particularly common for organic molecules having a permanent dipole, e.g., acetone. These molecules are frequently referred to as polar molecules, a use of the term polar which is very different than used in this book.

- Hydrogen bonding: The organic molecule can be either the donor or the acceptor or both, depending on the nature of the clay surface and the organic molecule.

- π Bonding: Molecules such as benzene can interact via their π electrons with, for example, Cu^{2+} interlayer cations.

Hydrogen bonding and π bonding are both examples of Lewis AB (electron-donor/electron-acceptor) interactions; see Chapters 5 and 8.

3.6.1 Organic complexes with vermiculite

Johns and Sen Gupta (1967) studied a series of complexes of *n*-alkyl ammonium cations with vermiculite. Depending on the treatment of the cation exchanged vermiculite, four different structures were found. All contained the straight chain cation with its -N-H_3 group keyed into the ditrigonal hole of the vermiculite layer surface and tilted at about 55° (±5°) from the normal. In the phase which contains only the organic cation (the so-called collapsed phase), the orientation of the alkyl chain is such that the C-N bond at the end of the chain is essentially perpendicular to the clay layer surface (Figure 3.15). This arrangement allows the three hydrogen atoms on the nitrogen to form hydrogen bonds to three of the six proximal oxygen atoms of the ditrigonal hole, thus anchoring the molecule both through hydrogen bonds and by way of the electrostatic attraction between the charged ammonium moiety and the tetrahedral charge caused by aluminum for silicon substitution. Note that adjacent chains are anchored to different vermiculite layers promoting the maximum contactable surface between adjacent alkyl chains. This arrangement promotes an LW attraction between adjacent chains. Further, the interlayer volume now has a portion which looks like an organic liquid (the alkyl chains), albeit a highly organized liquid.

The LW interaction between parallel alkyl chains is non-negligible (see Chapter 8). Given the ability of the interlayer organic cations to attract, via the LW forces, other hydrocarbon moieties, it is not surprising that exposure of the *n*-alkyl ammonium vermiculite complex to the equivalent chain length amine results in a more complex two-layer organoclay (Figure 3.17). The two organic layers are similarly oriented with the nitrogen adjacent to the vermiculite layer surface. As such, half of the molecules have, as in the single layer structure described above, the N-C bond perpendicular to the surface allowing hydrogen bonding to the surface oxygen atoms. These, by analogy to the single layer structure, are the *n*-alkyl ammonium cations. The remaining molecules have a different orientation with a much more inclined N-C bond. These presumably are the uncharged *n*-alkyl amine molecules which are not forming hydrogen bonds to the vermiculite surface. Similar structures can be obtained by adding the same length *n*-alkyl alcohol to the organoclay (Lagaly and Weiss, 1969).

The other complexes found by Johns and Sen Gupta (1967) involved the addition of 1) water molecules, and 2) a molecule of the *n*-alkyl ammonium salt. These structure determinations were based on an analysis of the 1-dimensional X-ray diffraction intensities (the 00ℓ reflections) and yield only the distance (the z-coordinates) the carbon and nitrogen atoms from the silicate layer. Thus, there is no information about the lateral (the x- and y-coordinates) of the organic molecules and the clay layers.

More information about the interaction of organic cations with vermiculite comes from a study of the anilin-vermiculite complex (Slade and Stone, 1983) and the anilinium-vermiculite complex (Slade *et al.*, 1987). As in the *n*-alkyl ammonium complexes, the ammonium group keys into the ditrigonal holes of the vermiculite layer. The phenyl groups are oriented nearly perpendicular to the phyllosilicate layer and, in the ideal case, occupy all ditrigonal sites. This requires that adjacent phenyl groups be rotated by about 90°. The arrangement of one set of aniline cations is shown in Figure 3.14. The anilinium cations, for simplicity, are all shown with the nitrogen oriented toward one vermiculite surface. It is likely that the ammonium group is oriented toward both adjacent silicate surfaces, as is the case of the *n*-alkyl ammonium vermiculite structures (Johns and Sen Gupta, 1967). The disorder in the anilinium vermiculite complex contains both orientations but these are expressed in terms of an average structure.

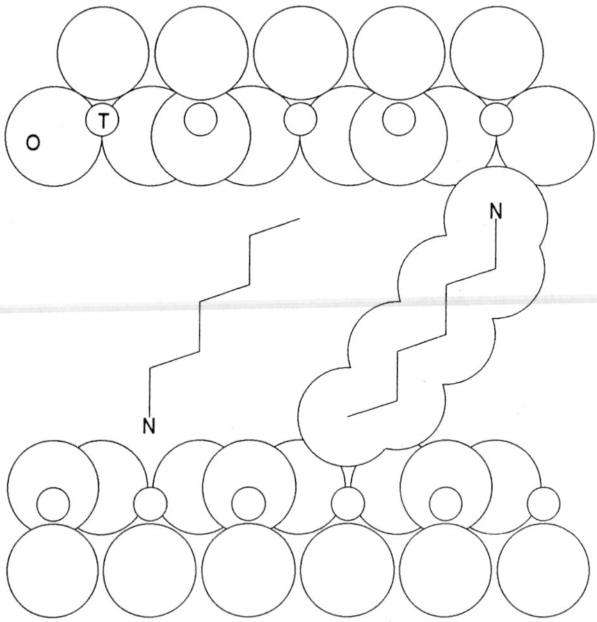

Figure 3.15: *The arrangement of the single layer of n-alkyl ammonium cations in a vermiculite (the collapsed phase). The "T" refers to the tetrahedral cations and "O" marks the oxygen atoms. After Johns and Sen Gupta (1967).*

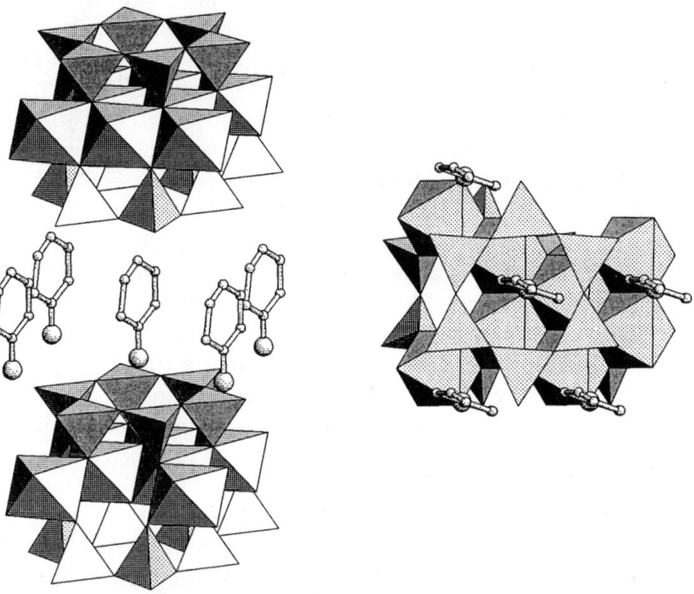

Figure 3.16: *The arrangement of aniline cations in the vermiculite complex. Only one of several orientations of the molecule are shown.*

These two studies of the structure of vermiculite with different organic cations show that for a high charge phyllosilicate, the ammonium group keys into the ditrigonal cavity of the silicate layer, driven by the electrostatic interaction between the cation and the electrostatic imbalance in the clay layer due to ionic substitution. If there are hydrogen atoms bonded to the nitrogen, these also contribute to the bonding between the organic cation and the clay layer. Presumably, replacement of the ammonium hydrogen atoms by small organic moieties, e.g., methyl groups, will not change significantly the manner in which the organic cations attach themselves to the clay layer. For lower charge phyllosilicates, e.g., montmorillonite, the density of charge on the silicate surface is reduced, there are fewer organic cations per unit area and the orientation of the alkyl chain can be quite different from the examples discussed above.

There is an obvious relation between the orientation of the organic cation and the layer charge of the phyllosilicate. The surface area of the phyllosilicate is, for dioctahedral minerals, about 23 Å per T_4O_{10}. The arrangement of the ditrigonal cavities, into which the polar group keys, is such that the distance between adjacent cavities is approximately 2.6 Å, with a surface coverage of approximately 23 $Å^2$ for an assumed circular contact. Work by Lagaly and Weiss (1969) shows that for a layer charge of 1 (a mica), the n-alkyl chains are approximately close packed with a tilt angle (the angle between the axis of the chain and the surface of the mineral) of about 90°. Assuming that as the layer charge is reduced in a continuous fashion, the average distance between the n-alkyl ammonium chains increases and as a result, the chains will tilt to maintain close physical contact. Under these assumptions, it is possible do derive a relation between the tilt angle and layer charge. The tilt angle varies (Figure 3.17) in an approximately linear manner for layer charges less than 0.8 and is sharply curved for layer charges between 0.8 and 1.0. Also shown on the plot are experimental values of Lagaly and Weiss (1969). This simple model assumes that for all layer charges examined, the length of the n-alkyl chain exceeds the distance between the polar groups which are attached to the clay surface. If the chains are shorter, then the chain will adopt an orientation with is parallel to the clay surface (see Figure 1 in Lagaly and Weiss (1969)). It should be emphasized that the whole discussion in this section has dealt with information derived from X-ray diffraction studies of organic molecules situated between the layers of phyllosilicates. The placement of the organic molecules is determined by the chemical and physical properties of the surfaces of the phyllosilicate layers, and the orientation of the molecules is strongly influenced by the existence of the molecule between two silicate layers. When we examine the situation

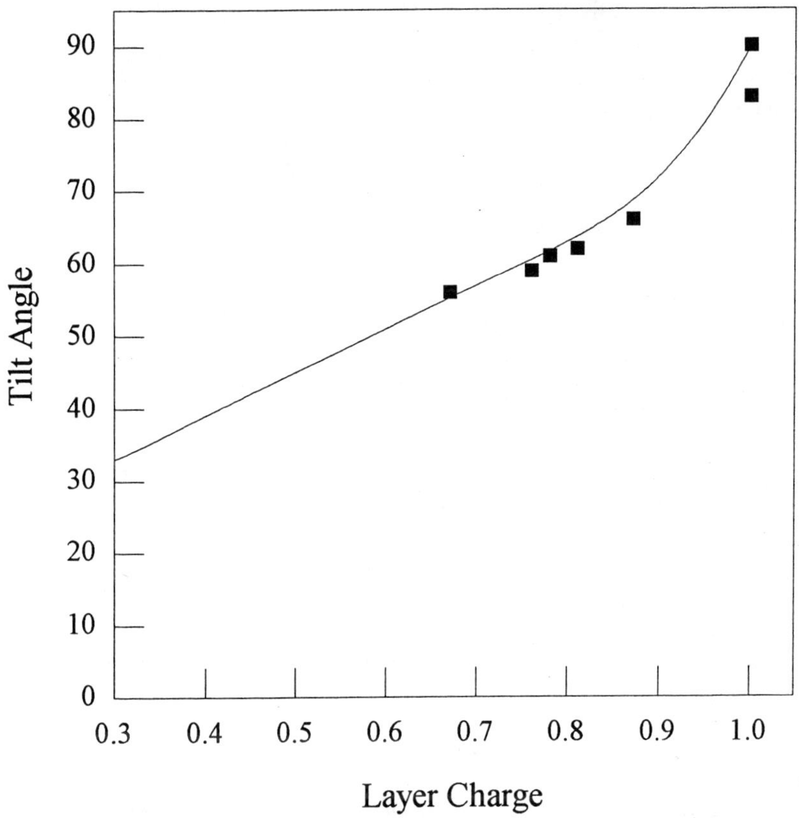

Figure 3.17: *The relation between tilt angle and layer charge for n-alkyl ammonium cations. The line is calculated from simple assumptions detailed in the text; the points are from Lagaly and Weiss (1969).*

of the same molecules attached to the *external* surfaces of the same mineral, the chemical and physical properties of the external layer remain essentially unchanged from that of a layer buried in the crystal, but there is no bounding layer on the other side. In addition, the molecule attached to the external surface can be in contact with a gas or liquids of different kinds.

3.6.2 Organic complexes with kaolinite

The structure of the layer in kaolinite is very different from that of smectites and vermiculites. In the latter, the 2:1 structures, the interlayer void is bounded on both sides by essentially the same surface; basal oxygen atoms of the tetrahedra. In contrast, the 1:1 structures, of which kaolinite is the most common, have interlayer voids bounded by, on one side the same basal oxygen atoms and on the other side by a closest packed arrangement of hydroxyl groups. There are other differences as well; with the exception of talc and pyrophyllite, there are exchangeable interlayer cations and water or other molecular species in the 2:1 minerals. Kaolinite has essentially no ionic substitution, therefore a zero layer charge and nothing between the layers. Thirdly, the forces holding the layers together in the 2:1 minerals (excepting talc and pyrophyllite) are electrostatic, LW forces and hydrogen bonding, especially when water is present (Giese and Fripiat, 1979). Since there is nothing to replace or to hydrate in the interlayer region of kaolinite, it is more difficult to intercalate materials in this mineral. In general, small highly polar organic molecules can intercalate kaolinite directly, by simple exposure of the clay to the liquid or vapor of the organic material. There are no single crystals of kaolinite so the techniques applied to vermiculite intercalates cannot be used. Recently, it has become possible to take X-ray and neutron diffraction data from powder samples and refine the crystal structure using the Rietveld approach of fitting a calculated powder diffraction scan to the observed intensities (Rietveld, 1969).

Dimethylsulfoxide (DMSO) easily forms an intercalate with kaolinite (Olejnik *et al.*, 1968). The crystal structure was determined from a combination of X-ray diffraction and neutron diffraction giving a very accurate structure and the location of all hydrogen atoms. The polar character of DMSO results from the asymmetry of the molecule which has two methyl groups on one side of the sulfur and an oxygen on the other side. The dipole is aligned between the oxygen atom and the two methyl groups with the oxygen end being the negative end of the dipole. The most obvious interaction between DMSO and the kaolinite layer is between the oxygen atom and the surface hydroxyls, allowing hydrogen bonding, and this is the geometrical arrangement in the intercalate (Figure 3.18). Because kaolinite is dioctahedral, there is a missing octahedral cation and this creates a modest cavity in the hydroxyl surface, much like the ditrigonal hole of the tetrahedral sheet. The DMSO oxygen partly fits into this and is in close proximity to three of the surface hydroxyls. These hydroxyls are tilted toward the oxygen and that plus the relatively short O(H)-O(DMSO) distances are consistent with hydrogen bonding.

This orientation places one of the methyl groups close to the basal oxygen atoms of the adjacent layer. Hydrogen bonding from the methyl group to the oxygen atoms is unlikely so the factor governing the clay-molecule interaction at this end of the molecule is the keying into the ditrigonal hole of the uppermost methyl group. In the second study of interest here, the intercalating compound was N-methylformamide. This molecule has, as is the case for DMSO, the oxygen keyed into the vacant octahedral site of the hydroxyl sheet and receives three hydrogen bonds from that surface (Figure 3.19a). The C-O bond is roughly perpendicular to the hydroxyl surface, and this places the amide nitrogen close to the basal oxygens of the adjacent tetrahedral sheet. There is no hydrogen bonding from the amide nitrogen to these basal oxygens. Such an orientation places the hydrogen on the nitrogen in the center of the ditrigonal hole of the tetrahedral sheet (Figure 3.19b). The hydrogen atoms of the methyl group were not located in the structure study, suggesting that there is a rotational disorder resulting from a lack of specific interactions between these hydrogen atoms and the clay surfaces.

3.6.3 Summary of molecular-clay interactions

The previous discussion of the few detailed structures indicate that the interactions between the interlayer molecular species and the adjacent clay surfaces are of two general types: specific interactions and non-specific interactions. The examples cited above give two instances of specific interactions. The hydrogen bonding from a 1:1 hydroxyl surface to a hydrogen bond acceptor where the donor is a Lewis acid and the acceptor is a Lewis base, as in the kaolinite-DMSO and dickite-N-methylformamide complexes is one kind. (In general the terminology is: electron-acceptors are Lewis acids and electron-donors are Lewis bases.) A second example is the positioning of the positively charged ammonium head of an *n*-alkyl ammonium cation close to the source of the negative charge of a 2:1 structure. This positioning may also allow hydrogen bonding from the NH_3 group to the surface oxygen atoms which reinforces the attachment. Where there is no specific interaction, as in the case of the nitrogen of N-methylformamide in proximity to

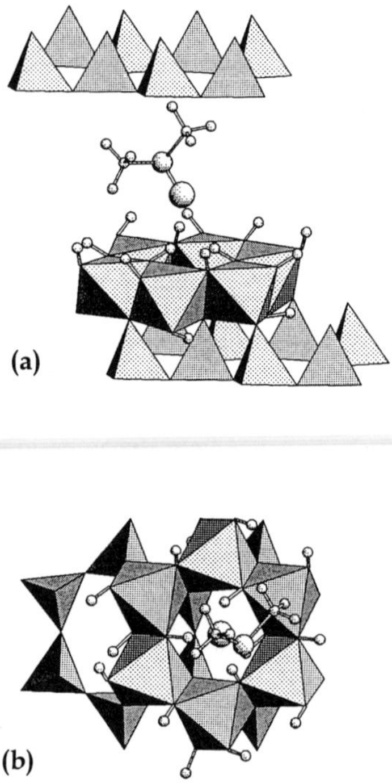

Figure 3.18: *Two views of the DMSO-kaolinite intercalate. In a) the keying of the oxygen atom above the vacant octahedral site is shown, as well as the tilting of the surface hydroxyl groups toward the oxygen. The view in b) shows the orientation of the DMSO molecule and the manner in which the uppermost (in this view) methyl group fits into the ditrigonal hole of the adjacent 1:1 layer.*

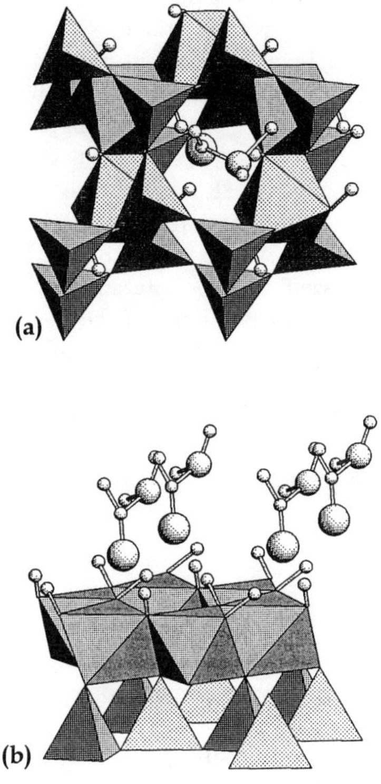

Figure 3.19: *The arrangement of the N-methylformamide molecule intercalated in dickite. View a) is a clinographic perspective of the molecules in contact with the hydroxyl surface of the dickite layer. The oxygen atom keys into the vacant octahedral site. View b) shows the placement of a single molecule between two adjacent layers with the N-H part of the molecule uppermost.*

the basal oxygen atoms and the alkyl chains of n-alkyl ammonium cations to each other and to the phyllosilicate surfaces, the tendency is to arrange the molecular species to maximize molecular contacts while minimizing the volume occupied by the interlayer material.

We have less detailed knowledge about the interactions of water and molecular species with interlayer cations. There are more possibilities for placement and orientation of these species leading to a level of disorder that obviates structure determinations. Even so, the same sorts of general specific and non-specific interactions also govern the physical placement of the molecular species. A specific interaction would be that between an interlayer cation (a Lewis acid) and e.g., an oxygen atom (a Lewis base) of ethylene glycol. Another example would be the interaction between Cu^{2+} and the π electrons of benzene, again an interaction of the Lewis acid-base type.

Non-specific interactions are typified by the orientation of the alkyl chains of organic cations either along the oxygen surface of a 2:1 phyllosilicate or directed away from the oxygen surface so that the chains may interdigitate.

All of the interactions discussed here are of a non-covalent type. As such they can be treated as interfacial interactions using the formalism described in Chapter 5.

3.7 Origin of Clay Minerals

A detailed discussion of the origin of mineral colloids is far beyond the scope of this book. This section follows the review paper of Eberl (1984).

Typically, silicate minerals form at some depth in the Earth where the ambient pressure and temperature are elevated. These minerals then either form directly from molten silicate material (magma) or precipitate from hot water (hydrothermal). In contrast, clay minerals, largely form at very low temperatures and pressures, often by mechanisms that are influenced by their layer structures. As such, it is worthwhile discussing the most common means of forming clay minerals.

3.7.1 Modes and environments of formation

The occurrence of a particular clay mineral is ultimately governed by thermodynamics which depends on the chemical environment of the locality where the mineral is formed. There are basically three mechanisms for the formation of clay minerals. The production of a mineral directly from a solution of appropriate ions is termed *neoformation*. The deposition of clay minerals which formed elsewhere (by neoformation, for example) is called *inheritance.* And the alteration of a pre-existing clay mineral to form another mineral while keeping the original layer structure is *layer transformation.* The layer transformation may simply involve the exchange of the interlayer and surface bound ions or it may result in major changes in the chemistry and perhaps the structure of the clay layers themselves.

The conditions under which these mechanisms involve combinations of temperature, pressure and chemical environment. Rather than try to specify exact conditions, it is useful to describe environments which include a range of such conditions. Thus, one environment is that in which *weathering* is the dominant process. This occurs at or very near the surface of the Earth, in contact with the atmosphere and water is available either as ground water or water infiltrating from rain or rivers. Thus, the temperatures are moderate and the amount of (fresh) water available varies considerably depending on the local climate. The *sedimentary* environment refers to the accumulation of minerals, largely fine-grained, at the sediment-water interface. This commonly occurs in lakes and the ocean. For ocean deposition, the most common situation, temperatures are lower and more restricted than for the weathering environment. Hydrostatic pressures range from 10^5 to 10^8 Pa. The chemistry of the water varies from that of normal sea water to quite different compositions as the sea water in the sediment pores reacts with the minerals present. At slightly elevated temperatures and pressures, there is the *diagenetic/hydrothermal* environment describes conditions where the minerals are in contact with hot water. Finally, if talc and pyrophyllite are to be included in the group of clay minerals, higher temperatures and pressures are needed and we have the *metamorphic* environment. In principal, the combination of the three mechanisms and the four environments yield twelve combinations for the production of clay minerals.

Out of these many combinations, a few generalizations emerge. For example, variation in rainfall in a warm climate where weathering is fairly rapid will produce clay minerals based largely on the solubility of the constituents of these minerals. The least soluble mineral in a typical geological environment containing alkali, aluminum, silicon and water would be gibbsite (Al_2O_3) the most soluble clay mineral is a smectite, for example $Na_{0.3}Al_{1.7}Mg_{0.3}Si{-}4O_{10}(OH)_2$, with kaolinite, $Al_2Si_2O_5(OH)_4$ having an intermediate solubility. So in a region with high rainfall, weathering would produce a soil rich in gibbsite while a drier but otherwise comparable situation would have a smectite in the soil.

In the oceans, where temperatures are low and reaction rates are slow, inheritance is the major process for the deposition of clay minerals. The clays originate from weathering on the adjacent continents and are transported to the oceans by rivers and the wind. About the only change that occurs after the deposition of the continental clays in the oceans is ion exchange whereby any calcium in exchange positions is replaced by sodium and magnesium, reflecting the composition of sea water.

Temperatures and pressures increase as sediment is buried, providing energy for further changes in the clay minerals. Shales are particularly rich in (inherited) clay minerals, and much effort has gone into understanding how smectite minerals change into illite during burial. This phenomenon was pointed out by Burst and Powers (Burst, 1959; Powers, 1967). Originally, the transformation was thought to involve the replacement of tetrahedral silicon by aluminum, thus increasing the layer charge. At the same time, potassium which is released by the dissolution of K-feldspar exchanges for the smectite interlayer cation. Eventually the layer charge increases to the point that the interlayer potassium dehydrates and the end product is illite (Perry and Hower, 1970; Hower *et al.*, 1976). Subsequent work has suggested several other mechanisms which are admirably discussed by Moore and Reynolds (1997). The transformation is an important one because it occurs under conditions where primitive organic material in the shale is converted to hydrocarbons.

3.7.2 Commercial deposits of clay minerals

Most clay minerals are dispersed in sedimentary rocks and are not of economic interest. Special conditions of formation and accumulation can produce deposits which are particularly rich in a single clay mineral. Clay minerals typically are low-cost items so that any clay deposit has to be at or near the surface of the ground where it can be inexpensively mined by powered shovels and bulldozers. There are probably as many special circumstances leading to a clay deposit as there are different clay minerals. Two examples suffice to illustrate these origins.

Smectite deposits are largely the result of the alteration of beds of volcanic glass. Certain types of volcanic eruption produce large volumes of glass and these deposits can be several meters thick. Frequently, the ash accumulates in shallow marine basins where the glass alters to smectites. These can produce important deposits of smectite as are found in Wyoming (Crook County; SWy-1), Arizona (Apache County; SAz-1), and Texas (Gonzales County; STx-1).

Kaolinite occurs in a different geological situation. The older deposits in Cornwall, UK and Georgia, USA resulted from weathering or mild hydrothermal conditions of feldspar grains in older igneous rocks. In the Georgia deposits, the kaolinite was transported from the weathering site to a number of small basins by running water. Thus, these are secondary deposits. In contrast, the Cornwall deposits have not been transported and the kaolinite remains interspersed in the weathered igneous rock. The weathering has destroyed the mechanical integrity of the parent rock. The separation of the kaolinite is accomplished by high pressure water hoses which break up the weathered rock, allowing the small particles of kaolinite to be washed into a basin at the bottom of the open pit.

4

Other Mineral Colloids

There are some 3,600 mineral species presently recognized (Fleischer and Mandarino, 1995). The vast majority of these are rare or occur only in isolated deposits. As such, they are of little importance in a geological sense. The common and geologically important minerals occur in two settings; in the rocks forming the surface and near surface of the Earth and those occurring in soils and sediments (Table 4.1).

These are the materials which humans frequently encounter. If one restricts the discussion, as in this book, to those common materials which occur either naturally or as the result of manufacturing processes as colloids, the list becomes very much shorter. Even with this shorter list, there are relatively few minerals whose surface thermodynamic properties have been determined. This reflects the fact that the underlying theory of surface thermodynamic components is relatively new and there are presently few researchers active in studying the properties of minerals as opposed to polymers and biological materials, for example.

In this section, the minerals whose surface thermodynamic properties have been measured are described in terms of their structure, chemistry and, where possible, the atomic arrangement of the external surfaces of finely divided particles. Since this chapter is not meant to be a detailed description of the structures of these materials, much of the normal information such as unit cell parameters, space group and atomic positional and thermal parameters are not listed. These data are easily located in the literature using the references provided. Good general accounts of these minerals are provided by Klein and Hurlbut (1993) and, in more detail, by Putnis (1992).

As noted in Chapter 2, the oxide minerals (including silicate minerals) are frequently composed of large oxygen anions and very much smaller cations of the first three rows of the periodic table. This allows the oxygen anions

Mineral Group	Percentage
Plagioclase	39%
Alkali Feldspars	12
Quartz	12
Pyroxenes	11
Amphiboles	5
Micas	5
Clay Minerals	5
Other Silicates	3
Non-silicates	8

Table 4.1: *The mineral composition of the combined oceanic and continental crust, expressed in volume % (Klein and Hurlbut, 1993).*

to assume a closest packed arrangement with the smaller cations in the interstices. Closest packing is the arrangement of equal-sized spheres which occupies the minimum volume. The concept of closest packing is very old, having been treated by Keplar in 1611 (Singh, 1997). The question of packing equal spheres in two dimensions has a unique solution. The pattern is shown in Figure 4.1. Each sphere has six spheres in contact with it forming an hexagonal pattern. Addition of a sphere on top of this two-dimensional array is easily achieved by placing the next sphere in contact with three spheres of the first layer (Figure 4.1). It turns out that there are two different sites which allow this; the easiest way to distinguish between these is to look at the space between the triad of spheres of the first layer. One site (Figure 4.1) has the shape of a triangle with an apex pointed up (labeled 2 in the figure) and the other site (labeled 1 in the figure) has the triangle oriented with the apex pointed down.

To add a second closest packed layer to the first, all the spheres in the second layer must occupy the same kind of site, either 1 or 2 but not a mixture as in Figure 4.1. Spheres occupying a mixture of 1 and 2 sites cannot form a complete layer because the distance between the two sites is smaller than the diameter of a sphere. Thus, at some point on the second level, spheres will physically prevent occupying two adjacent 1 and 2 sites, breaking the continuity of the second layer. We now have two layers in a closest packed arrangement. Adding a third (and further) layers introduces complications. This comes about because there are two distinct sites available for spheres in the third layer (Figure 4.2). One site (labeled 1 in Figure 4.2) lies directly

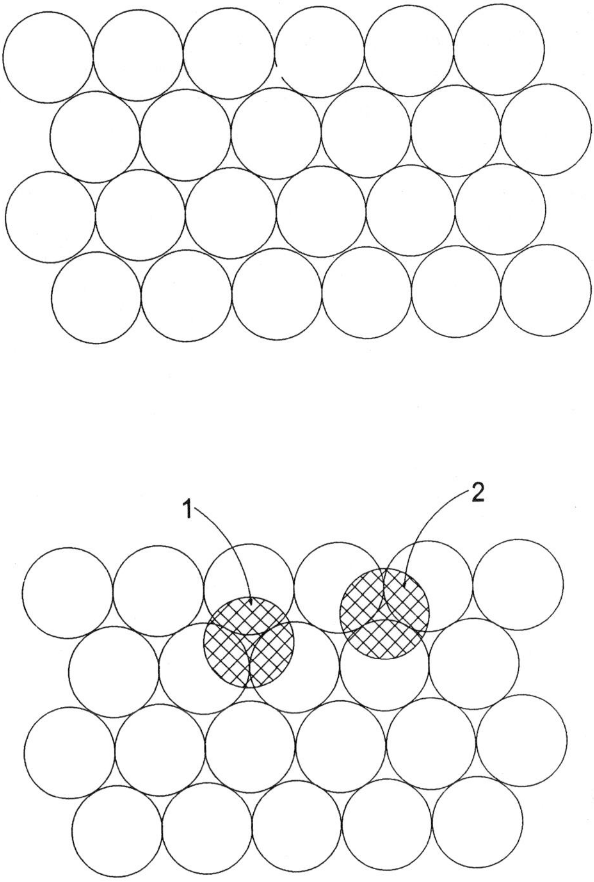

Figure 4.1: *A closest packed layer of equal sized spheres (above) and the same layer with two isolated spheres (shaded) added to a second layer. The added spheres, labeled 1 and 2, occupy different sites on the surface of the first layer.*

above a vacant space in the first layer while the other site (labeled 2 in Figure 4.2) lies directly above a sphere in the first layer. If, as often done, the spheres of the first layer are labeled "A", then the sphere in the second layer is labeled "B" (it does not superpose on A), and the sphere in the third layer is *either* "C" (labeled 1 in Figure 4.2; it does not superpose on either A or B) *or* "A" (2 in Figure 4.2; it superposes on A in the first layer). Thus, we have the stacking of closest packed layers in an ABC pattern or an AB pattern. Repeating these patterns as ...ABCABCABC... or ...ABABAB... yields two configurations; *cubic closest packing* or *ccp* and *hexagonal closest packing* or *hcp*. The distinction between these two cases only exists when there are more than two layers of closest packed spheres. The names for the two stackings derive from the symmetry of the arrangement of equal sized spheres. The mineral equivalent of these closest packed arrangements are the oxides where the cations are either tetrahedrally or octahedrally (or both) coordinated. These minerals range from simple oxides to carbonates, various forms of silica and the layer silicates (see Chapter 3).

There are a number of compendia of mineral properties and structures (e.g., Smyth and Bish (1988)). The data for the following descriptions have largely been taken from the Inorganic Crystal Structure Database published by the Gmelin-Institut fur Anorganische Chemie, FIZ Karlsruhe. Each structure from this database is identified by the ICSD number. For descriptions taken from other sources, the appropriate citation is given. For each mineral, interatomic distances are listed from the cations to oxygen atoms (or other anions where appropriate). If there are several identical distances, a single value is listed followed by a numerical superscript indicating the number of such distances, e.g., 1.456^4 Å would indicate that there are four identical distances of 1.456 Å.

4.1 Simple Oxides

Simple oxides have the general formula AX with the divalent metal cation occupying octahedral sites in a cubic closest packed arrangement of oxygen atoms. The parent compound, not an oxide, is the mineral halite, NaCl, which has a cubic closest packing of anions. Periclase, MgO, Figure 4.3, has the same atomic arrangement as halite with oxygen occupying the chloride positions and magnesium replacing the sodium. Hematite (Fe_2O_3) and corundum (Al_2O_3) have the same structure, based on an hexagonal closest packing (Figure 4.4). Both have the trivalent cation in octahedral sites. One slightly unusual feature is that the octahedra share edges in a two dimensional array. In a phyllosilicate, this arrangement would be termed dioctahedral (see

Figure 4.2: *Two closest packed layers of equal sized spheres showing two distinct sites where spheres for a third layer can be placed (above). The site labeled 1 when occupied yield a cubic closest packing while occupancy of the site labeled 2 corresponds to hexagonal closest packing. The interstitial sites in closest packing are octahedra and tetrahedra (below).*

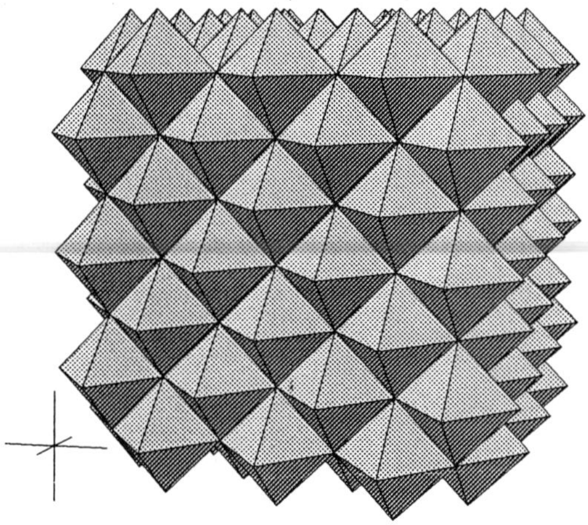

Figure 4.3: *A clinographic view of the periclase structure showing the octahedral coordination of oxygen atoms about the magnesium cations. The density of the mineral is 3.56 g/cc.*

Figure 4.4: *A clinographic view of the hematite and corundum structures showing the octahedral coordination of oxygen atoms about the aluminum cations, in the case of corundum or ferric cations, in the case of hematite. The density of hematite is 5.26 g/cc and 4.02 for corundum.*

Chapter 3) and only 2/3 of the octahedra are filled by cations. These arrays are stacked along the c-axis direction. Necessarily, some octahedra in adjacent octahedral planes share faces. This brings some of the trivalent cations close to each other which is unusual. Titanium forms a number of oxides only three of which occur as minerals; all have the composition TiO_2. The most common mineral is rutile (Figure 4.5) and the less common minerals are brookite (Figure 4.6) and anatase (Figure 4.7). The major differences between these three structures is the manner in which the octahedra share edges. The highest density mineral is rutile with D = 4.27 g/cm^3 and chains of octahedra share (two) edges. These chains are aligned along the c-axis and adjacent chains share corners. In brookite, three edges of each octahedron are shared with neighboring octahedra (D = 4.13 g/cm^3) while in anatase,

Central Atom	Ligand	Distances (Å)
Titanium (Rutile)	Oxygen	
tetragonal (P4 2/m n m)		1.942^4
		1.975^2
Titanium (Brookite)	Oxygen	
orthorhombic (P b c a)		1.864
		1.923
		1.931
		1.989
		1.999
		2.052
Titanium (Anatase)	Oxygen	
tetragonal (I 4_1/a m d)		1.934^4
		1.980^2

Table 4.2: *Titanium dioxide interatomic distances.*

each octahedron shares four of its edges (D = 3.89 g/cm^3).

Magnetite, Fe_3O_4, has iron in two sites; two thirds of the iron is in octahedral coordination with the remaining one third in tetrahedral coordination (Figure 4.8). This is the inverse spinel structure which has half the trivalent iron in tetrahedral coordination and the remaining half of the trivalent iron plus the divalent iron distributed in the octahedral sites. This can be described by the modified formula $^{IV}Fe^{3+}(^{VI}Fe^{2+}\,^{VI}Fe^{3+})O_4$ where IV refers to tetrahedral coordination and VI refers to octahedral coordination. The oxygen atoms are arranged in cubic closest packing.

The common ZrO_2 mineral is baddeleyite which is monoclinic (Figure 4.9). There are two high temperature forms which have higher symmetry; tetrag-

Central Atom	Ligand	Distance (Å)
Iron	Oxygen	
		1.889^4
Iron	Oxygen	
		2.058^6

Isometric (F d $\bar{3}$ m)

Table 4.3: *Magnetite interatomic distances.*

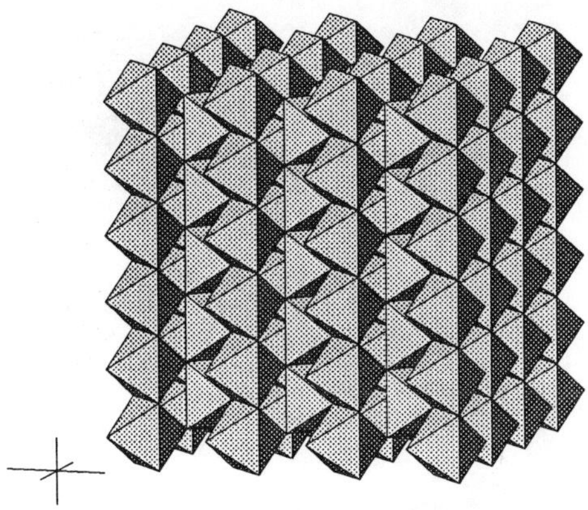

Figure 4.5: *A clinographic view of the rutile (TiO_2) structure.*

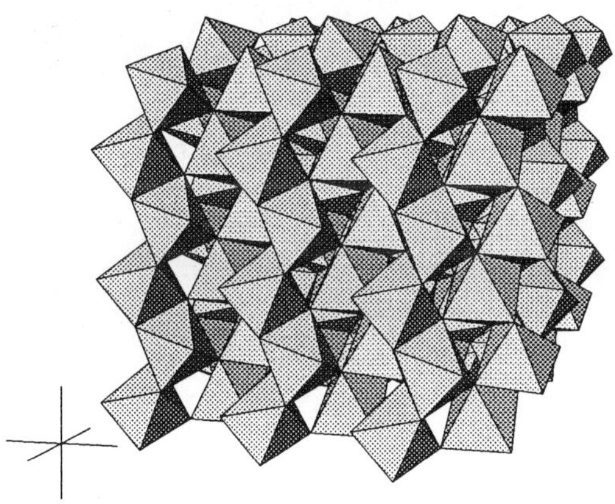

Figure 4.6: *A clinographic view of the brookite (TiO_2) structure.*

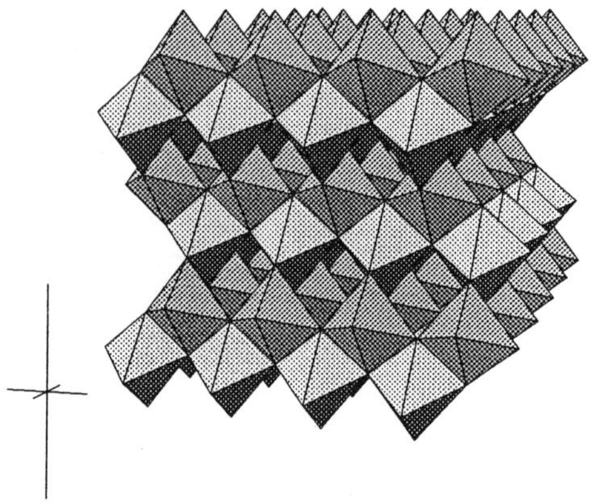

Figure 4.7: *A clinographic view of the anatase (TiO_2) structures.*

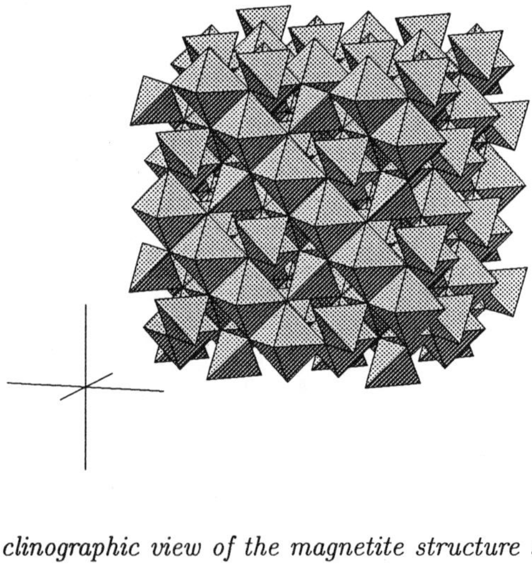

Figure 4.8: *A clinographic view of the magnetite structure showing the two kinds of coordination polyhedra; tetrahedral and octahedral. See the text for details.*

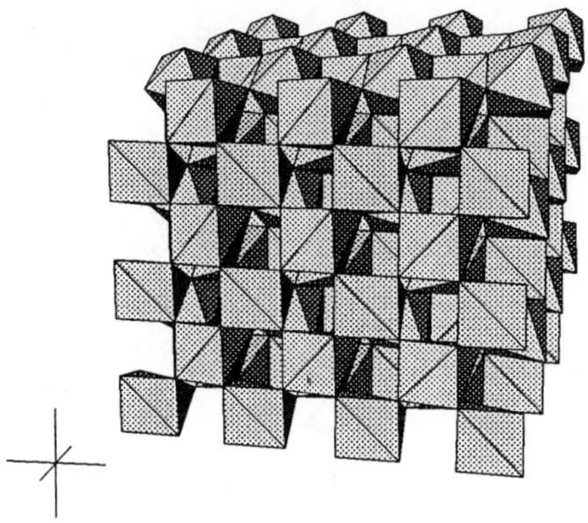

Figure 4.9: *A clinographic view of the baddeleyite structure showing the 7-coordinated oxygen polyhedra situated about each zirconium cation. The oxygen atoms are bonded to either 3 or 4 zirconium atoms. The ideal density is 5.83 g/cc.*

onal above about 1100°C and isometric above 2300°C. Small amounts of impurities (e.g., Mg) allow the higher symmetry forms to exist at room temperature. The monoclinic structure has seven oxygen atoms coordinating the zirconium (Table 4.4).

4.2 Halides

Many halides (e.g., halite (NaCl) and sylvite (KCl)) are very soluble and thus do not readily lend themselves to contact angle measurements with water. Thus, most of the surface thermodynamic measurements have been made on the insoluble halides such as fluorite (CaF_2). Fluorite has a structure with an unusual coordination of fluorine about the calcium; the fluorine atoms are arrayed at the corners of a cube with the calcium at the center (Figure 4.10). Each fluorine has four calcium atoms arranged in a tetrahedral manner (Table 4.5). The structure has an excellent octahedral cleavage. Many fluorite samples are strongly colored resulting from impurities.

Central Atom	Ligand	Distance (Å)
Zirconium	Oxygen	2.051
		2.057
		2.163
		2.151
		2.189
		2.285
		2.219
Oxygen	Zirconium	2.051
		2.057
		2.163
Oxygen	Zirconium	2.151
		2.189
		2.219
		2.285

Isometric (F d $\bar{3}$ m)

Table 4.4: *Baddeleyite interatomic distances.*

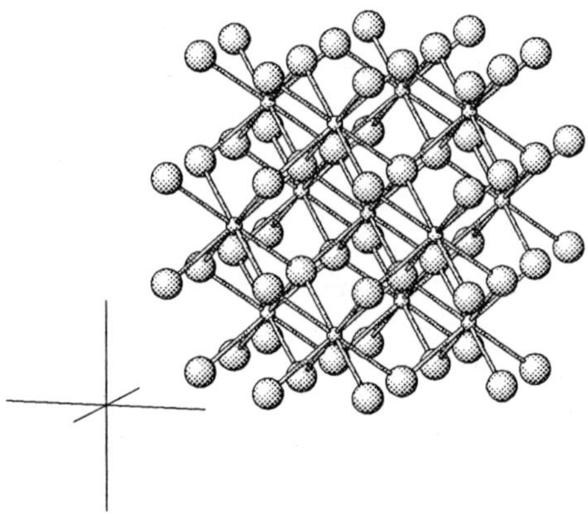

Figure 4.10: *A clinographic view of the fluorite structure. This view shows individual atoms, because the structure is obscured in the polyhedral mode. The ideal density is 3.18 g/cc.*

Central Atom	Ligand	Distance (Å)
Calcium	Fluorine	2.365^8
Fluorine	Calcium	2.365^4

Isometric (F m $\bar{3}$ m)

Table 4.5: *Fluorite interatomic distances.*

4.3 Hydroxides

There are two structure types in this group of minerals; the pure hydroxides and the oxyhydroxides. The atomic arrangement of the hydroxides is very simple. These are typified by brucite ($Mg(OH)_2$) and gibbsite ($Al(OH)_3$) (Figure 4.3). Both are layer structures where the layer is a simple planar arrangement of edge sharing octahedra. In the case of brucite, all the octahedral sites are occupied by divalent magnesium while in gibbsite, two thirds of the sites are filled by trivalent aluminum. Adjacent octahedral layers are joined by hydrogen bonds (Figure 4.11). The top and bottom surfaces of the octahedral sheets are formed by hydroxyl groups. In gibbsite, each hydroxyl is coordinated by three divalent cations in the sheet so the orientation of the hydroxyl bond is essentially vertical to the sheet. In contrast, the gibbsite hydroxyls have only two ligands within the octahedral sheet so there is freedom of orientation. Part of the hydroxyls lie in the plane of the octahedra while the remainder are nearly vertical, and form hydrogen bonds to the adjacent sheet.

4.4 Nesosilicates

The garnet group of minerals is large; there are 14 minerals listed in the Glossary of Mineral Species for 1995 (Fleischer and Mandarino, 1995). The general formula is $A_3B_2(SiO_4)_3$ where the A site is occupied by larger divalent cations (e.g., Ca, Mg, Fe^{2+}) and the B site is occupied by trivalent cations (e.g., Al, Fe^{3+} and Cr). The larger divalent cations occupy an 8-coordinated site, the trivalent cations are octahedrally coordinated and silicon is in a tetrahedral site (Figure 4.12). The interactomic distances are for an almandine sample (ideally $Fe_3Al_2(SiO_4)_3$) (Table 4.7).

Olivine is a name given to a group of minerals whose composition lies between the end members Mg_2SiO_4 and $Fe_2^{2+}SiO_4$. These compositions are

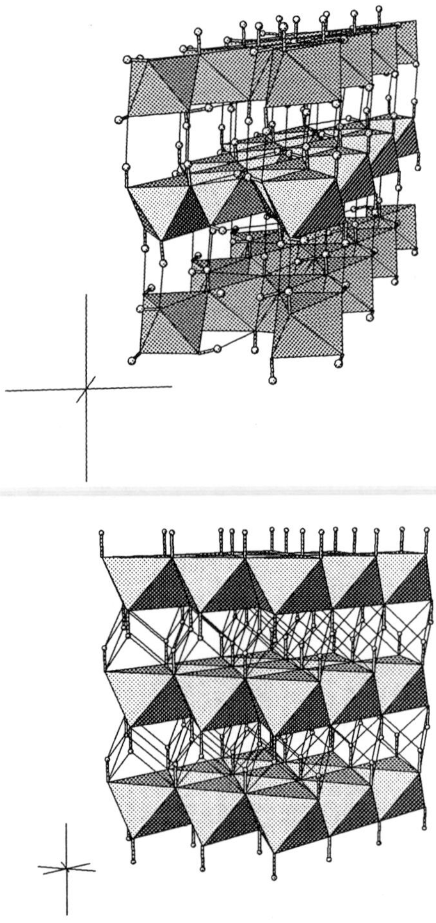

Figure 4.11: *A clinographic view of the gibbsite structure (above) and the brucite structure (below). The views show the coordination octahedra and the individual hydrogen atom (small spheres) with the hydroxyl bond shown as a stick. Interlayer hydrogen bonds are indicated by thin lines from the hydrogen atoms to oxygen atoms. The ideal density for brucite is 2.38 g/cc and for gibbsite is 2.42 g/cc.*

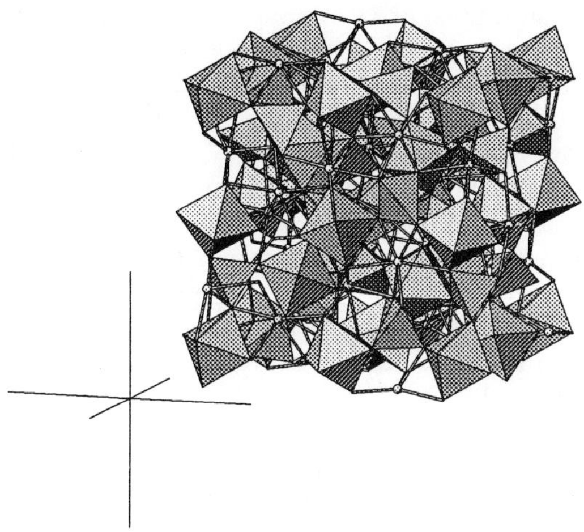

Figure 4.12: *A clinographic view of the garnet (almandine) structure. The structure is shown as octahedra and tetrahedra for the trivalent and tetravalent cations, but the higher coordination number of eight is shown as the central cation with bonds to the neighboring oxygen atoms. The ideal density for almandine is 4.31.*

typically greenish is color, hence the name. The divalent cations are octahedrally coordinated while the silicon is tetrahedrally coordinated (Figure 4.13). In the olivine structure there are two different magnesium atoms each in an octahedral site but the symmetry of each is different.

4.5 Cyclosilicates

The best known member of this group of silicates is the mineral beryl, $Be_3Al_2Si_6O_{18}$, which is known as emerald when it is colored a dark green. Another mineral, tourmaline, is fairly common in pegmatites and is prized for its varying colors. It is a semiprecious stone. The structure and chemistry are complex. The general formula is $WX_3Y_6(BO_3)_3Si_6O_{18}(OH)_4$. Here W is an alkali metal (K or Na) or an alkaline earth (Ca), X is a number of atoms such as aluminum, iron and magnesium, Y is a trivalent cation such as aluminum and iron. The basic structure is based on rings of silica tetrahedra with the composition Si_6O_{18}. These are held together by octahedra of the X variety and a higher coordination (9) with a W cation in the center. Boron occurs as triangular arrangements of oxygen atoms coordinating boron (Figure 4.14; Table 4.9).

Central Atom	Ligand	Distance (Å)
Gibbsite		
Aluminum	Oxygen	1.832
		1.906
		1.910
		1.918
		1.922
		1.925
Monoclinic ($P2_1/n$)		
Brucite		
Aluminum	Oxygen	2.099^6
Trigonal ($P\,\bar{3}\,m$)		

Table 4.6: *Hydroxide interatomic distances.*

Central Atom	Ligand	Distance (Å)
Iron (A site)	Oxygen	2.220^4
		2.378^4
Aluminum (B site)	Oxygen	1.896^6
Silicon	Oxygen	2.099^6

Isometric (I a $\bar{3}$ d)

Table 4.7: *Almandine (garnet) interatomic distances.*

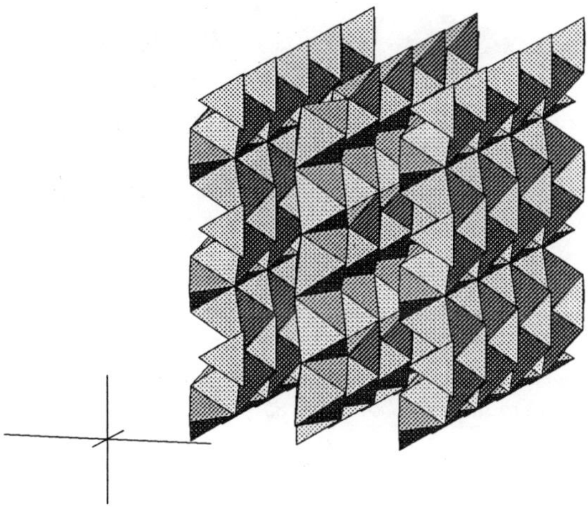

Figure 4.13: *A clinographic view of the forsterite (olivine) structure. The ideal density for forsterite is 3.23 g/cc.*

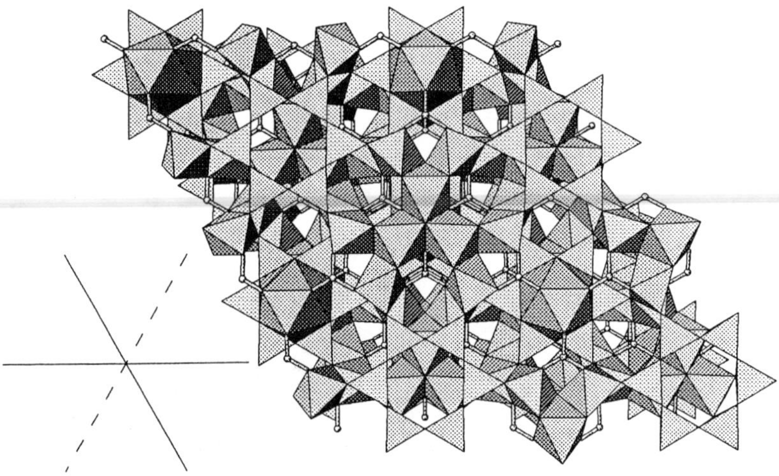

Figure 4.14: *A projection of the tourmaline (dravite) structure. The range of densities for tourmaline is 3.00 to 3.25 g/cc.*

4.6 Sorosilicates

These are not very common minerals, and the only example included here is vesuvianite. This is a complex structure with both isolated silica tetrahedra and sorosilicate groups. The ideal formula is $Ca_{19}Al_4Fe(Al,Mg,Fe)_8Si_{18}O_{70}(OH)_8$. The calcium is eight coordinated, aluminum is six coordinated, the disordered site is six coordinated and iron is both six and five coordinated. There are disordered sites in the structure which complicates the minerals. The interatomic distances are shown in Table 4.10 and the structure is shown in Figure 4.15.

4.7 Pyroxenes

These are composed of single chains of corner sharing tetrahedra which are stacked in pairs (see Chapter 2). Each chain, to a good approximation, has the tetrahedral bases (and their oxygen atoms) lying in a plane with apical oxygens lying above the plane. The pairs of chains are inverted so that the apical oxygen atoms are adjacent to each other and these form part of the coordinating oxygen atoms about the non-tetrahedral cations. These chains are bonded together by two type of (usually) divalent cations. These sites are labelled M1 and M2. The M1 sites are smaller and a rather regular octahedral sites. In contrast, M2 sites are less regular and are larger. When bothe M1 and M2 are occupied by smaller cations such as magnesium, the resulting structure yields the *orthopyroxenes* (orthorhombic symmetry)

Central Atom	Ligand	Distance (Å)
Magnesium(1)	Oxygen	2.067^2
		2.085^2
		2.132^2
Magnesium(2)	Oxygen	2.051
		2.068^2
		2.182
		2.213^2
Orthorhombic (P b n m)		

Table 4.8: *Forsterite interatomic distances. There are two independent magnesium atoms in the structure, although each has essentially the same environment.*

Central Atom	Ligand	Distance (Å)
Magnesium	Oxygen	1.953
		1.989^2
		2.002^2
		2.117
Sodium	Oxygen	2.504^3
		2.746^3
		2.842^3
Aluminum	Oxygen	1.900
		1.908
		1.915
		1.931
		1.960
	(OH)	2.002
Boron	Oxygen	1.373
		1.374^2
Silicon	Oxygen	1.605
		1.606
		1.625
		1.645
Trigonal (R 3 m)		

Table 4.9: *Tourmaline interatomic distances; the data are from a sample of dravite which is a sodium- magnesium variety. The 9- 6- and 4-coordinated groups are shown as polyhedra with the boron shown as individual atoms. In this view, the six tetrahedra are shown with their bases toward the viewer.*

Central Atom	Ligand	Dist. (Å)	Central Atom	Ligand	Dist. (Å)
Silicon (1)	Oxygen	1.635^4	Silicon (2)	Oxygen	1.608
					1.640
					1.644
					1.674
Silicon (3)	Oxygen	1.599	Calcium (1)	Oxygen	2.314^4
		1.622			2.521^4
		1.633			
		1.661[a]			
Calcium (2)	Oxygen	2.317	Calcium (3)	Oxygen	2.358
		2.341			2.451
		2.363			2.458
		2.417		(OH)	2.481
		2.423			2.492
		2.450			2.551
		2.482			2.574
		2.893			2.609
Aluminum (AlFe)	Oxygen	1.896	A-site (ideally Al)	Oxygen	1.862^2
		1.907		(OH)	1.873^2
	(OH)	1.925			1.944^2
		1.944			
		1.959			
		2.054			
B-site (ideally 0.5 Fe)	Oxygen	2.086^4	C-site (ideally 0.5 Ca)	Oxygen	2.296^4
		2.245			2.646^4
Tetragonal (P 4/n n c)					

Table 4.10: *Vesuvianite interatomic distances.* [a] *indicates the bridging oxygen between two Si(3) cations forming the* Si_2O_7 *group.*

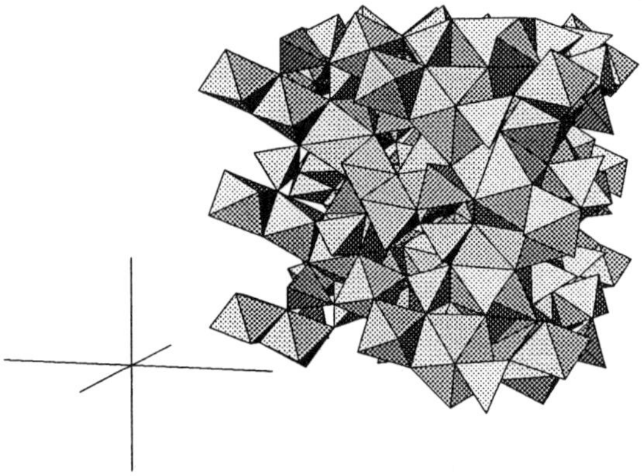

Figure 4.15: *A clinographic view of the vesuvianite structure. The ideal density for vesuvianite is 3.23 g/cc.*

while when a larger cation such as calcium is in the M2 site we have the *clinopyroxenes* (monoclinic symmetry). The examples presented here are enstatite ($MgSiO_3$, an orthopyroxene; Figure 4.16, Table 4.11) and diopside ($CaMgSi_2O_6$, a clinopyroxene; Figure 4.17, Table 4.12).

4.8 Amphiboles

As with the pyroxene minerals, there are two structure types for the amphibole minerals. These are double chains, so they stack in a similar manner to the single chain pyroxene minerals. The presence of larger cations, e.g., calcium, as in tremolite (Figure 4.17, Table 4.12)($Ca_2Mg_5Si_8O_{22}(OH)_2$), favor the monoclinic structure while smaller cations, e.g., magnesium, as in anthophyllite (Figure 4.18, Table 4.13) ($Mg_7Si_8O_{22}(OH)_2$), favor the orthorhombic structure. The double chains of tetrahedra form a string of (ideally) hexagonal rings. One of the consequences of this is that an oxygen atom which coordinates some of the divalent cations is underbonded since it is situated in the center of hexagonal ring of tetrahedra and therefore does not bond to a tetrahedral cation. The underbonding is resolved by adding a hydrogen atom to form a hydroxyl group. This same is true of the phyllosilicate minerals.

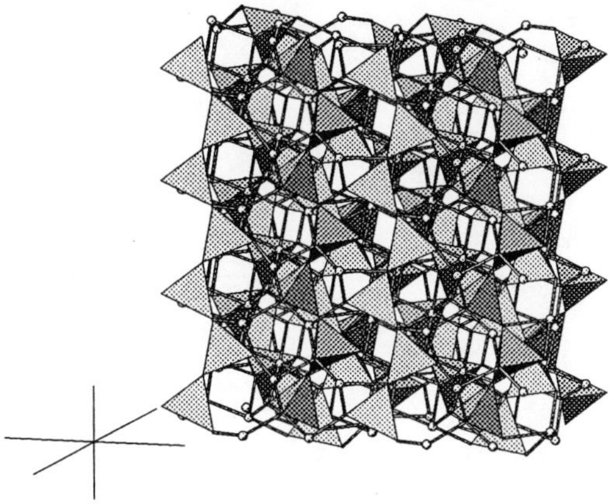

Figure 4.16: *A clinographic view of the enstatite structure, an orthopyroxene. The ideal density for enstatite is 3.20 g/cc.*

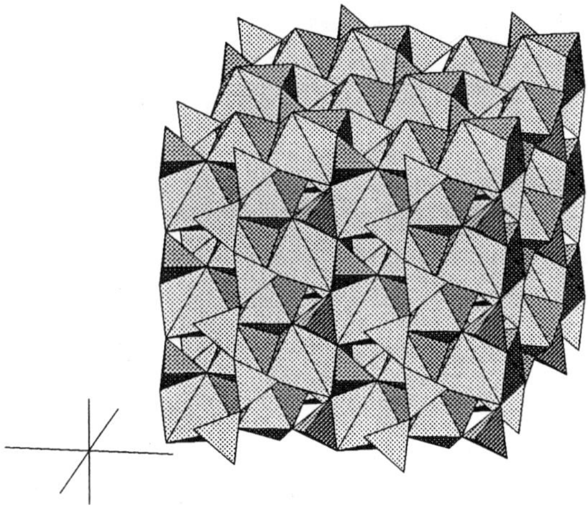

Figure 4.17: *A clinographic view of the diopside structure, a clinopyroxene. The ideal density for diopside is 3.28 g/cc.*

Central Atom	Ligand	Distance (Å)
Magnesium (M1 site)	Oxygen	2.007[4]
		2.028
		2.045
		2.045
		2.065
		2.151
		2.172
Magnesium (M2 site)	Oxygen	1.993
		2.032
		2.056
		2.088
		2.288
		2.448
Silicon	Oxygen	1.587
		1.611
		1.646
		1.665
Silicon	Oxygen	1.588
		1.618
		1.675
		1.678
Orthorhombic (P b c a)		

Table 4.11: *Enstatite, an orthopyroxene, interatomic distances.*

Central Atom	Ligand	Distance (Å)
Calcium (M2 site)	Oxygen	2.352^2
		2.360^2
		2.561^2
		2.717^2
Magnesium (M1 site)	Oxygen	2.050^2
		2.064^2
		2.115^2
Silicon	Oxygen	1.584
		1.602
		1.664
		1.687
Monoclinic (C 2/c)		

Table 4.12: *Diopside, a clinopyroxene, interatomic distances.*

The greater width of the chain in the amphibole minerals means that there are more types of cation sites for the non-tetrahedral cations. These are, as for the pyroxenes, labeled as M-sites. The scheme is that the structure can be split by a plane bisecting the center of the double chains. The M-site which lies on this plane is M3, the sites lying on just on either side of the plane are the M1 sites, the one lying further away are the M2 sites and furthest away and containing the calcium in the clinoamphiboles are the M4 sites.

4.9 Silica Minerals

The silica minerals might seem to be particularly simple, being constructed of SiO_4 tetrahedra which share all four corners with other SiO_4 tetrahedra. Yet the reality is that there are at least nine natural and synthetic polymorphs (Klein and Hurlbut, 1993). Part of this richness is due to the common occurrence of high- and low-temperature varieties (e.g., high-quartz which is stable above 574°C and low-quartz which is stable below that temperature). The change from the low to the high forms do not involve breaking bonds, there is an increase in symmetry and a decrease in density. In contrast, the change from e.g., quartz to tridymite involves breaking of strong Si-O bonds. As a result, these reconstructive transformations are usually sluggish and one frequently finds polymorphs existing well out of equilibrium. The examples of silica structures listed below are the low-temperature varieties.

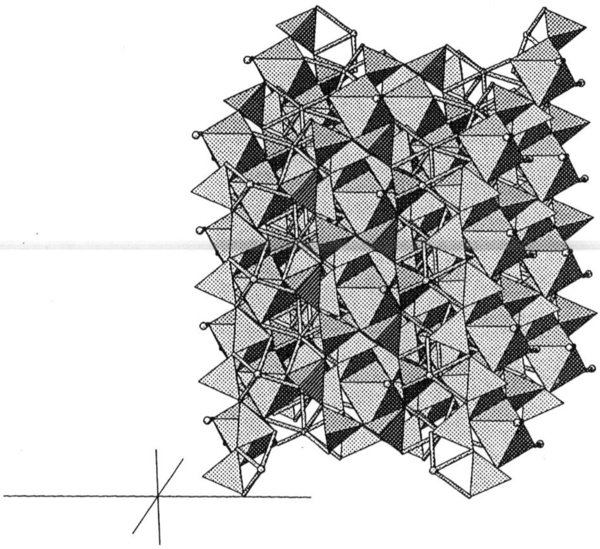

Figure 4.18: *A clinographic view of the tremolite structure, a clinopyroxene. The ideal density for tremolite is 2.90 g/cc.*

Central Atom	Ligand	Distance (Å)
Calcium (M4 site)	Oxygen	2.321^2
		2.396^2
		2.540^2
		2.767^2
Magnesium (M1 site)	Oxygen	2.064^2
		2.078^2
		2.083^2
Magnesium (M2 site)	Oxygen	2.014^2
		2.083^2
		2.134^2
Magnesium (M3 site)	Oxygen	2.057^2
	Oxygen	2.071^4
Silicon	Oxygen	1.587
		1.617
		1.655
		1.673
Silicon	Oxygen	1.587
		1.617
		1.655
		1.673

Monoclinic (C 2/m)

Table 4.13: *Tremolite, a clinoamphibole, interatomic distances.*

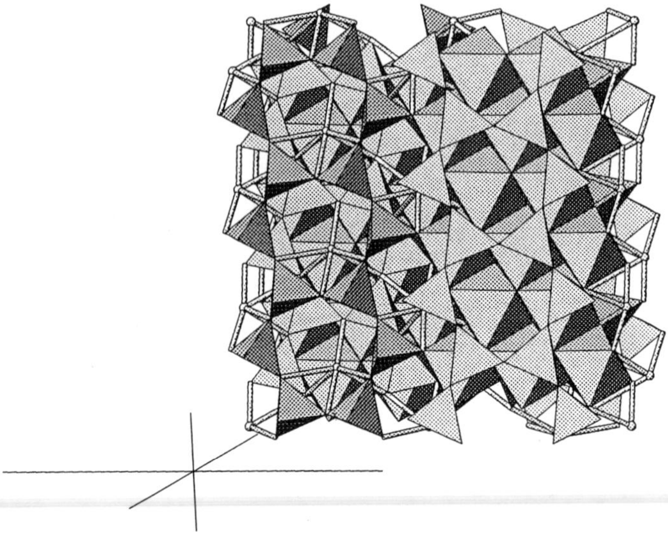

Figure 4.19: *A clinographic view of the gedrite structure, an orthoamphibole. The density for anthophyllite is about 3.2 g/cc.*

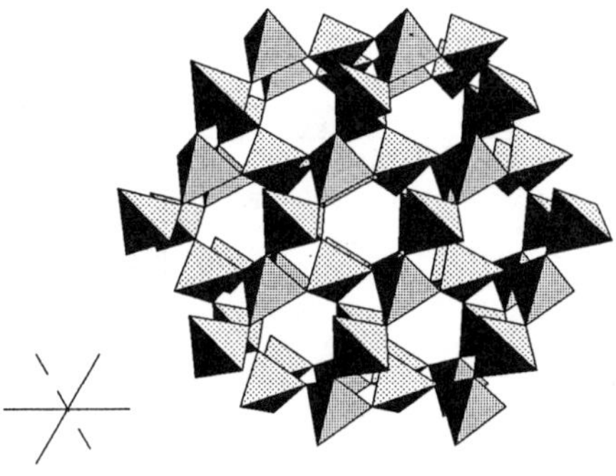

Figure 4.20: *A projection of the low-quartz structure along the trigonal axis. The density for quartz is about 2.6 g/cc.*

Central Atom	Ligand	Dist. (Å)	Central Atom	Ligand	Dist. (Å)
T1A	Oxygen	1.673	T1B	Oxygen	1.665
(.66Si		1.651	(0.62Si		1.658
.34Al)		1.640	0.38Al)		1.654
		1.635			1.643
T2A (Si)	Oxygen	1.638	T2B	Oxygen	1.670
		1.635	(Si)		1.648
		1.607			1.641
		1.579			1.630
M1	Oxygen	2.054	M2	Oxygen	1.924
(.12Fe		2.061	(.60 Al		1.951
.88Mg)		2.067	.36Mg .04Fe)		1.985
		2.078			2.005
		2.130			2.021
		2.158			2.028
M3	Oxygen	2.017	M4	Oxygen	2.015
(.90Mg .10Fe)		2.023	(.55Mg		2.103
		2.055^2	.42Fe .02Ca)		2.123
		2.097^2			2.217
					2.222
					2.416
A	Oxygen	2.30			
(.34Na		2.41			
.66vacant)		2.64^2			
		2.65^2			
Orthorhombic (P n m a)					

Table 4.14: *Gedrite, an orthoamphibole, interatomic distances.*

Central Atom	Ligand	Distance (Å)
Silicon	Oxygen	1.608
		1.608
		1.610
		1.610
Trigonal (P $3_2$21)		

Table 4.15: *Low temperature quartz interatomic distances.*

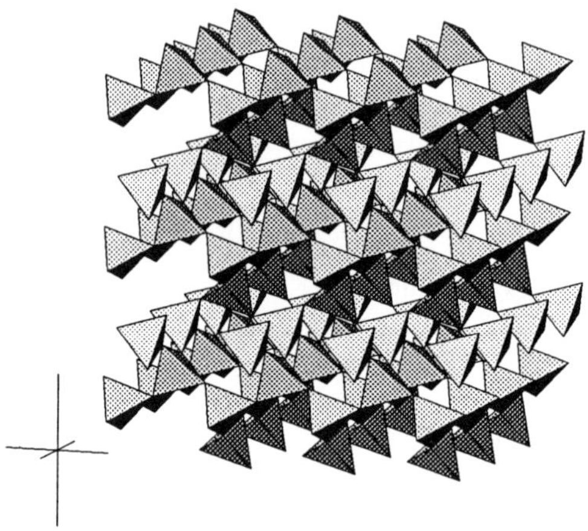

Figure 4.21: *A clinographic view of the cristobalite structure. The density for cristobalite is about 2.3 g/cc.*

Central Atom	Ligand	Distance (Å)
Silicon	Oxygen	1.605^2
		1.608^2
Tetragonal (P 4_12_12)		

Table 4.16: *Low-cristobalite interatomic distances.*

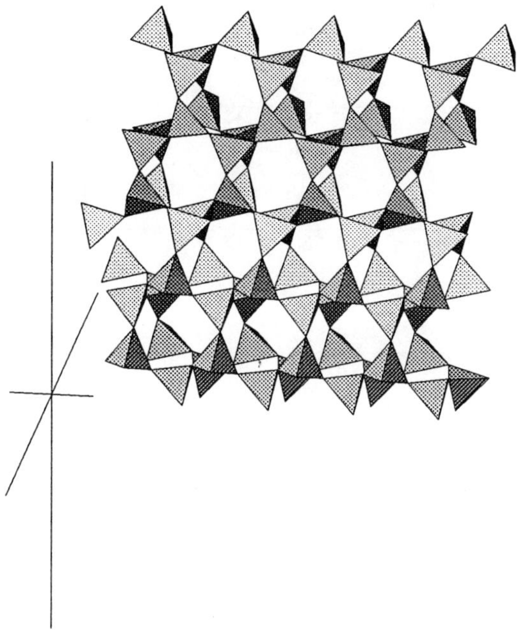

Figure 4.22: *A clinographic view of the low-tridymite structure. The density for tridymite is about 2.3 g/cc.*

Central Atom	Ligand	Dist. (Å)	Central Atom	Ligand	Dist. (Å)
Si (1)	Oxygen	1.576	Si (2)	Oxygen	1.566
		1.602			1.587
		1.606			1.607
		1.634			1.630
Si (3)	Oxygen	1.587	Si (4)	Oxygen	1.574
		1.596			1.577
		1.607			1.615
		1.608			1.620
Si (5)	Oxygen	1.568	Si (6)	Oxygen	1.551
		1.604			1.577
		1.604			1.604
		1.622			1.618
Si (7)	Oxygen	1.585	Si (8)	Oxygen	1.572
		1.592			1.585
		1.614			1.593
		1.627			1.609
Si (9)	Oxygen	1.565	Si (10)	Oxygen	1.575
		1.573			1.577
		1.590			1.586
		1.595			1.614
Si (11)	Oxygen	1.562			
		1.592			
		1.611			
		1.622			
Monoclinic (C c)					

Table 4.17: *Low temperature tridymite interatomic distances.*

4.10 Feldspars

The feldspars are an extremely complex group of minerals. Their chemistry is deceptively simple; the variation can be represented in terms of three end members. These are $KAlSi_3O_8$ nominally called orthoclase, $NaAlSi_3O_8$ nominally called albite and $CaAl_2Si_2O_8$ nominally called anorthite. They are related by simple substitution (orthoclase and albite are related by $K \rightleftharpoons Na$) or by a coupled substitution (albite and anorthite are related by $Na + Si \rightleftharpoons Ca + Al$). There is only limited exchange between orthoclase and anorthite. All these relations can be shown in a ternary diagram (Figure 4.26). The structures of these three minerals are based on the same architecture; both the aluminum and silicon atoms are tetrahedrally coordinated by oxygen and the larger alkali or alkaline earth cation fit into larger void where it is irregularly coordinated by oxygen. The complexities of real mineral samples lie in the arrangement of the aluminum and silicon atoms. For example, at very high temperatures where the structure is expanded and the atoms have more thermal energy, the aluminum and silicon will tend to distribute themselves in a random manner over the tetrahedral sites. If cooled rapidly, this arrangement will be preserved at room temperature. When viewed as a whole, a crystal of a specific feldspar, e.g., orthoclase, will appear to be composed of tetrahedra which are "average" and represent, in the case of orthoclase, 1/4 aluminum and 3/4 silicon. Individual tetrahedra of course contain either an aluminum or a silicon atom, but in a disordered system one cannot predict which atom will be in a specific tetrahedron. This atomic arrangement for disordered orthoclase (sanadine) is shown in Figure 4.23 and the interatomic disteances are listed in Table 4.18. The albite structure, as a comparison, represents an intermediate stage of ordering (or disordering). The structure is shown in Figure 4.24 and the interatomic distances are shown in Table 4.19.

4.11 Carbonates

The commonly found carbonate minerals are calcite ($CaCO_3$), dolomite ($CaMg(CO_3)_2$), and magnesite ($MgCO_3$). The former two minerals make up large volumes of carbonate rock, the ultimate repository for atmospheric carbon dioxide. These minerals have essentially the same crystal structure with rhombohedral symmetry which is characterized by excellent rhombohedral cleavages.

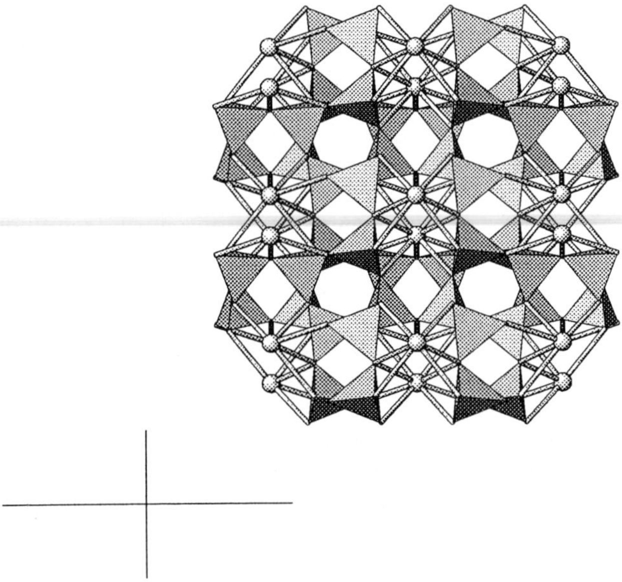

Figure 4.23: *A projection of the disordered orthoclase (sanadine) structure onto the a-b plane showing the vertical mirror planes. The density for orthoclase is about 2.6 g/cc.*

Central Atom	Ligand	Distance (Å)
T (0.25Al 0.75Si)	Oxygen	1.650
		1.652
		1.654
		1.657
T (0.25Al 0.75Si)	Oxygen	1.650
		1.652
		1.654
		1.657
T (0.25Al 0.75Si)	Oxygen	1.624
		1.634
		1.635
		1.639
T (0.25Al 0.75Si)	Oxygen	1.624
		1.634
		1.635
		1.639
Potassium	Oxygen	2.698
		2.898^{2}
		2.953^{2}
		3.029^{2}
		3.128^{2}
Monoclinic (C 2/m)		

Table 4.18: *Disordered orthoclase interatomic distances.*

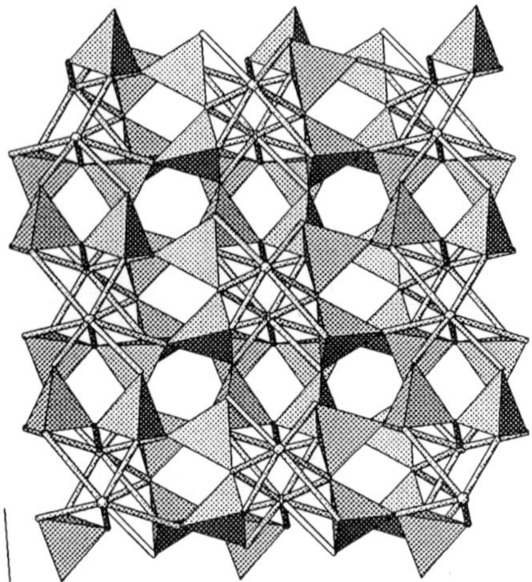

Figure 4.24: *A clinographic view of the albite structure. The density for albite is about 2.6 g/cc.*

4.12 Phosphates

Phosphate minerals are minor constituents of many type of rocks. The principal mineral is apatite $Ca_5(PO_4)_3OH$. This is a major constituent of teeth and bones as well as occurring in earth materials. Natural samples frequently exhibit substitution of the hydroxyl groups (in hydroxyapatite) by fluoride or chloride ions producing fluorapatite and chlorapatite, respectively.

4.13 Sulphates

The common sulphate minerals are gypsum $CaSO_4.2H_2O$ and anhydrite $CaSO_4$. These are frequently found in evaporatie deposits where they have been formed by the evaporation of sea water over long periods of time.

Central Atom	Ligand	Distance (Å)
T_1o	Oxygen	1.679
		1.676
		1.672
		1.682
T_1m	Oxygen	1.635
		1.619
		1.639
		1.632
T_2o	Oxygen	1.648
		1.627
		1.630
		1.624
T_2m	Oxygen	1.648
		1.625
		1.625
		1.635
Na	Oxygen	2.345
		2.462
		2.489
		2.626
		2.650
		3.014
		3.094
		3.245
		3.286

Triclinic ($C\,\bar{1}$)

Table 4.19: *Intermediate albite interatomic distances.*

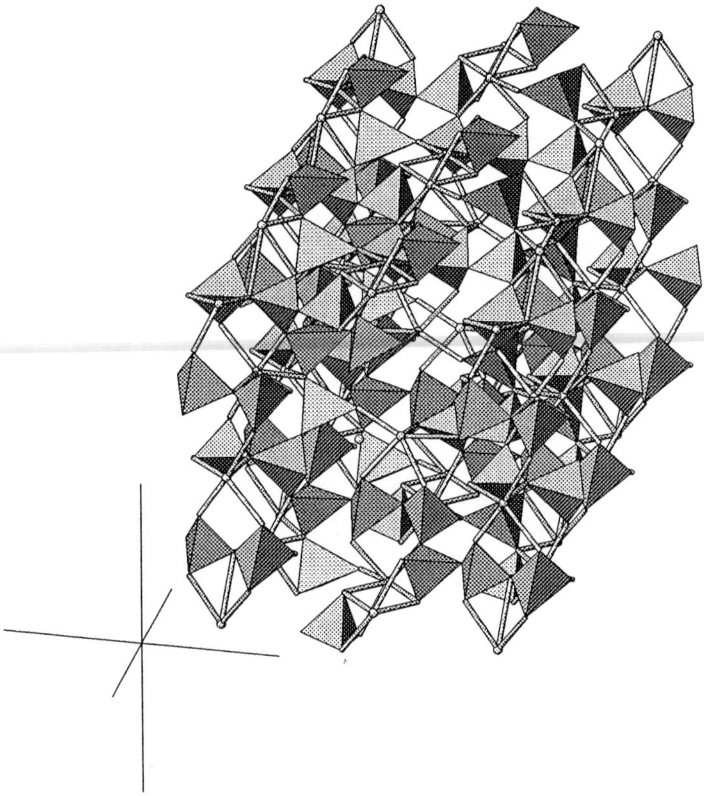

Figure 4.25: *A clinographic view of the anorthite structure. The density for anorthite is about 2.8 g/cc.*

Central Atom	Ligand	Dist. (Å)	Central Atom	Ligand	Dist. (Å)
Si (1)	Oxygen	1.535	Si (2)	Oxygen	1.568
		1.616			1.596
		1.632			1.962
		1.685			1.750
Si (3)	Oxygen	1.504	Si (4)	Oxygen	1.586
		1.589			1.624
		1.637			1.627
		1.732			1.755
Si (5)	Oxygen	1.590	Si (6)	Oxygen	1.591
		1.638			1.612
		1.640			1.631
		1.676			1.691
Si (7)	Oxygen	1.534	Si (8)	Oxygen	1.526
		1.569			1.631
		1.650			1.651
		1.662			1.736
Al (1)	Oxygen	1.632	Al (2)	Oxygen	1.652
		1.699			1.756
		1.722			1.827
		1.777			1.872
Al (3)	Oxygen	1.691	Al (4)	Oxygen	1.632
		1.721			1.725
		1.725			1.774
		1.811			1.792
Al (5)	Oxygen	1.585	Al (6)	Oxygen	1.678
		1.680			1.737
		1.717			1.767
		1.823			1.873
Al (7)	Oxygen	1.727	Al (8)	Oxygen	1.702
		1.744			1.726
		1.746			1.748
		1.751			1.750
Ca (1)	Oxygen	2.306	Ca (2)	Oxygen	2.362
		2.486			2.459
		2.497			2.615
		2.511			2.639
		2.570			2.643
		2.903			2.678
		3.110			2.733
Ca (3)	Oxygen	2.337	Ca (4)	Oxygen	2.097
		2.366			2.407
		2.403			2.462
		2.411			2.626
		2.680			2.678
		2.703			2.706
		2.717			2.827

Triclinic (P $\bar{1}$)

Table 4.20: *Anorthite interatomic distances. The distances about the Ca ions have been limited to 7 in number.*

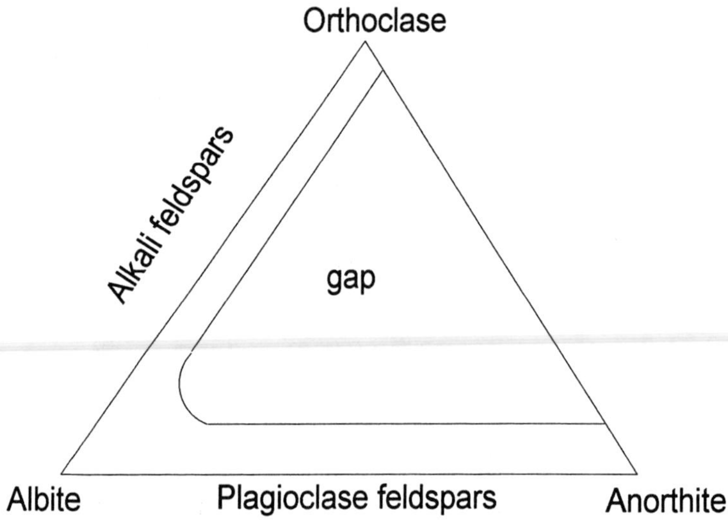

Figure 4.26: *The ternary diagram showing the observed variation in chemical composition in terms of the end members orthoclase, albite and anorthite. The extent of the stability field changes with temperature.*

Central Atom	Ligand	Distance (Å)
Ca	Oxygen	2.360^{6}
C		1.281^{3}
Rhombohedral (R $\bar{3}$c)		

Table 4.21: *Calcite interatomic distances.*

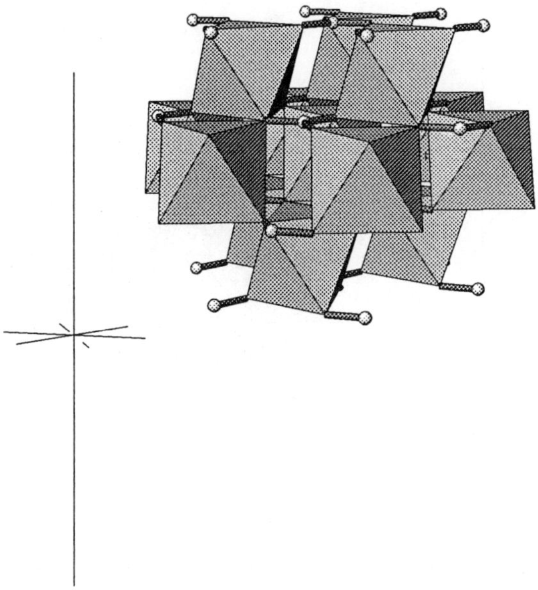

Figure 4.27: *A clinographic view of the calcite structure. The density for calcite is about 2.7 g/cc.*

Central Atom	Ligand	Distance (Å)
Ca	Oxygen	2.382^6
Mg	Oxygen	2.088^6
C	Oxygen	1.285^3
Rhombohedral (R $\bar{3}$)		

Table 4.22: *Dolomite interatomic distances.*

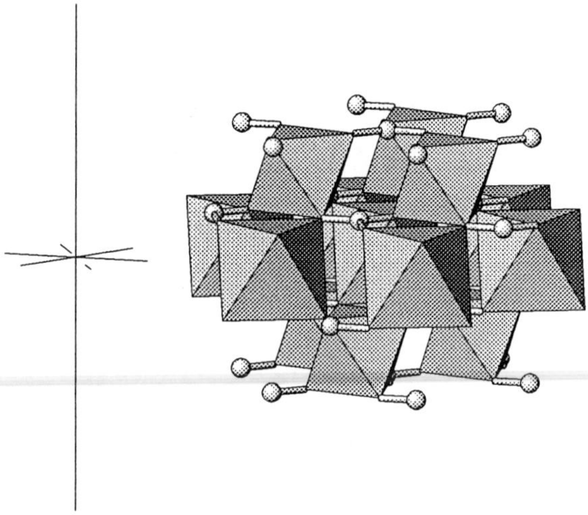

Figure 4.28: *A clinographic view of the dolomite structure. The density for dolomite is about 2.9 g/cc.*

Central Atom	Ligand	Distance (Å)
Mg	Oxygen	2.102^{6}
C	Oxygen	1.285^{3}
Rhombohedral (R $\bar{3}$c)		

Table 4.23: *Magnesite interatomic distances.*

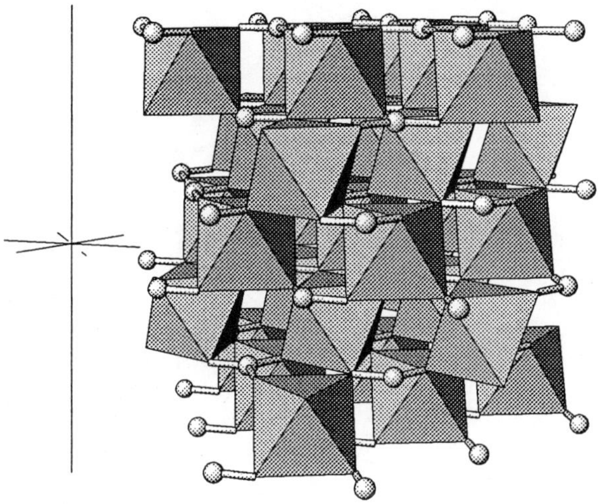

Figure 4.29: *A clinographic view of the magnesite structure. The density for magnesite is about 3.0 g/cc.*

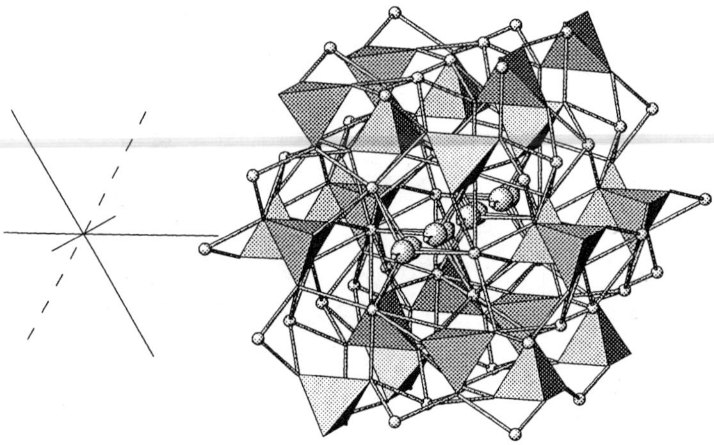

Figure 4.30: *A clinographic view of the hydroxyapatite structure. The density for hydroxyapatite is about 3.3 g/cc.*

Central Atom	Ligand	Distance (Å)
Ca (1)	Oxygen	2.407^2
		2.408
		2.454^2
		2.455
		2.807
		2.808^2
Ca (2)	Oxygen	2.346^2
		2.358
		2.383^2
		2.513^2
		2.708
P	Oxygen	1.529^2
		1.537
		1.547
H	Oxygen	0.957
Hexagonal (P 6_3/m)		

Table 4.24: *Hydroxyapatite interatomic distances.*

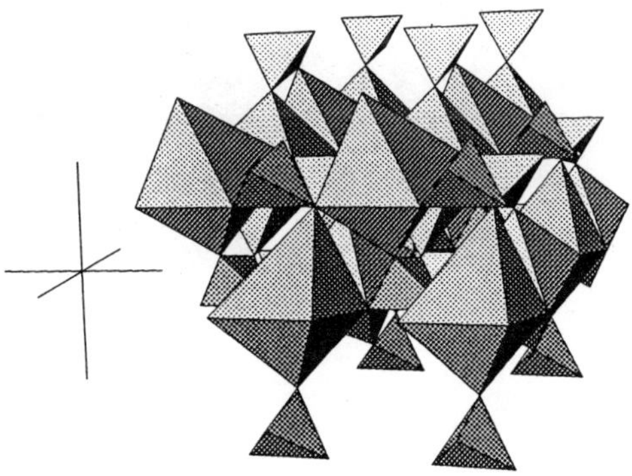

Figure 4.31: *A clinographic view of the barite structure. The density for barite is about 4.5 g/cc.*

Central Atom	Ligand	Distance (Å)
Ba	Oxygen	2.766
		2.808^2
		2.815^2
		2.911^2
S	Oxygen	1.454
		1.478
		1.485^2
Tetragonal (P 4_12_12)		

Table 4.25: *Barite interatomic distances.*

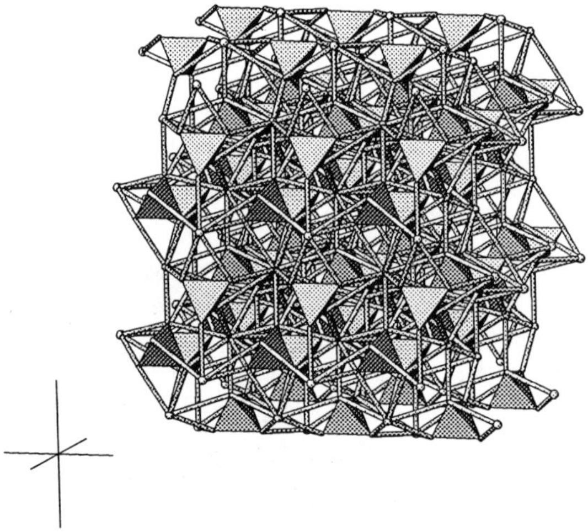

Figure 4.32: *A clinographic view of the celestite structure. The density for celestite is about 4.0 g/cc.*

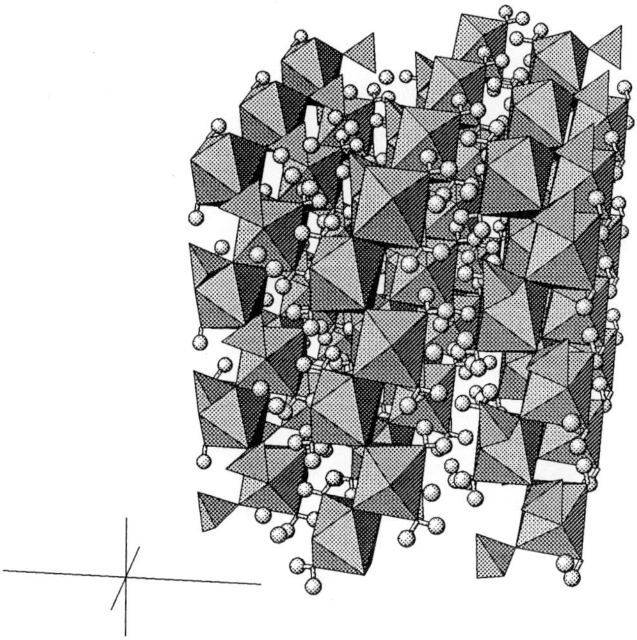

Figure 4.33: *A clinographic view of the gypsum structure. The density for gypsum is about 2.3 g/cc.*

Central Atom	Ligand	Distance (Å)
Ca	Oxygen	2.366^{2}
		2.374^{2}
		2.546^{2}
		2.552^{2}
S	Oxygen	1.471^{2}
		1.474^{2}
H	Oxygen	0.959
H	Oxygen	0.942
Monoclinic (I 2/c)		

Table 4.26: *Gypsum interatomic distances.*

Central Atom	Ligand	Distance (Å)
Ca	Oxygen	2.347^{2}
		2.464^{2}
		2.510^{2}
		2.563^{2}
S	Oxygen	1.472^{2}
		1.473^{2}
Orthorhombic (A mma)		

Table 4.27: *Anhydrite interatomic distances.*

4.14 Asbestos

The term "asbestos" is not a mineral name; rather it refers to a specific morphology or habit (termed "asbestiform") of several different minerals. The morphology is that of fibers whose aspect ratio (the ratio of the fiber length to the fiber diameter) is on the order of 100:1 or greater. The fibrous habit occurs in two very different ways: in one, the mineral crystal breaks along pronounced cleavage surfaces producing long fibers, and in the other, a platy mineral occurs not as flat sheets but as coiled tubes having a fibrous character. The amphibole minerals have the appropriate cleavage for a fibrous habit while the 1:1 phyllosilicate mineral chrysotile forms tubes by rolling up the silicate layers.

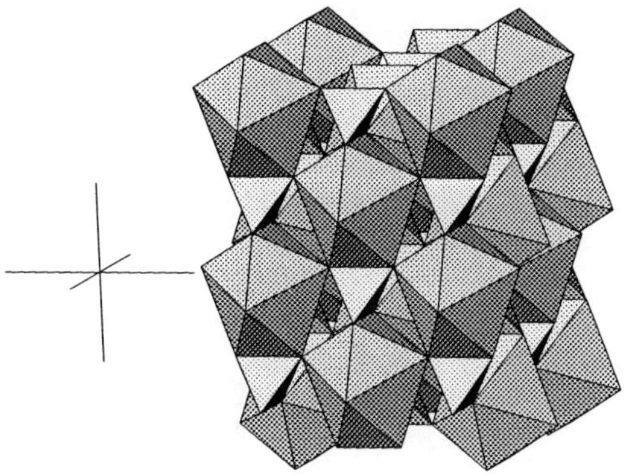

Figure 4.34: *A clinographic view of the anhydrite structure. The density for anhydrite is about 2.3 g/cc.*

5

Theory of Colloids

The first hint that there are non-covalent interactions between uncharged atoms and molecules came from the observations of van der Waals (1873, 1881). These interactions came to be known as "van der Waals forces." The interactions responsible for these became clear with the work of Keesom (1915, 1920, 1921), Debye (1920, 1921) and London (1930) as, respectively, interactions between two permanent dipoles (orientation forces), a permanent dipole and an induced dipole (induction forces) and a fluctuating dipole and an induced dipole (dispersion forces). While these three kinds of interaction have different origins, the interaction energies for all three vary as the inverse of the distance raised to the sixth power:

$$V_{\text{orientation}} = -\mu^4/kT\ell^6 \tag{5.1}$$

$$V_{\text{induction}} = -\alpha\mu^2/\ell^6 \tag{5.2}$$

$$V_{\text{dispersion}} = -\frac{3}{4}\,\alpha^2 h\nu/\ell^6 \tag{5.3}$$

where μ is the dipole moment, k is the Boltzmann constant, T is the absolute temperature, α is the polarizability, ℓ is the interatomic distance, ν is the main dispersion frequency and h is Planck's constant.

5.1 The Hamaker Approximation

For two atoms or molecules (i) separated by a short distance *in vacuo*, the dispersion interaction can be expressed as:

$$A_{ii} = \pi^2 q_i^2 \beta_{ii} \tag{5.4}$$

where A_{ii} is commonly termed the Hamaker constant, q_i^2 is the number of atoms per unit volume and $\beta_{ii} = 3/4\alpha^2 h\nu$. From Eq. 5.3, it follows that:

$$V_{ii} = -\beta_{ii}/\ell^6 \tag{5.5}$$

where ℓ is the distance between the atoms i. Extending the interaction to two different atoms i and j:

$$\beta_{ij} = \sqrt{\beta_{ii}\beta_{jj}} \tag{5.6}$$

It follows then that the Hamaker constant for two different atoms is given by:

$$A_{ij} = \sqrt{A_{ii}A_{jj}} \tag{5.7}$$

Hamaker (1937) first calculated the dispersion (van der Waals-London) interaction energy *for larger bodies* by a pair-wise summation of the properties of the individual molecules (assuming these properties to be additive, and non-retarded). Using this macroscopic approximation, the total dispersion energy for two semi-infinite flat parallel bodies (of material i), separated by a distance ℓ, in air or *in vacuo*, becomes (for ℓ greater than a few atomic diameters):

$$V_{\text{dispersion}} = -A_{ii}\ /12\pi\ell^2 \tag{5.8}$$

where A is the Hamaker constant for material i. From Eq. 5.3, A can therefore be expressed as:

$$A_{ii} = \pi^2 q_i^2 \beta_{ii} \tag{5.3a}$$

When two or more different materials interact, the total Hamaker constant is determined by a geometric combining rule (cf. Eq. 5.7). For two atoms of the same material, 1, in medium 3 (e.g., two individual clay particles in an aqueous suspension) the combining rule gives:

$$A_{131} = \left(\sqrt{A_{11}} - \sqrt{A_{33}}\right)^2 \tag{5.9}$$

or:

$$A_{131} = A_{11} + A_{33} - 2A_{13} \tag{5.10}$$

Here, the convention is that the first and the third character in the triplet subscript identify the two particles which are interacting through a liquid medium, identified by the second character. For two different objects (1 and 2), in medium 3:

$$A_{132} = A_{12} + A_{33} - A_{13} - A_{23} \tag{5.11}$$

Equation 5.11 can also be written as:

$$A_{132} = \left(\sqrt{A_{11}} - \sqrt{A_{33}}\right)\left(\sqrt{A_{22}} - \sqrt{A_{33}}\right) \tag{5.11a}$$

For the case where two objects of the same material are embedded in another material, the Hamaker constant, A_{131} 5.10, always is positive (or zero), for two different materials, the Hamaker constant, A_{132} 5.11, can be negative, i.e., when:

$$A_{11} > A_{33} > A_{22} \tag{5.12}$$

and when:

$$A_{11} < A_{33} < A_{22} \tag{5.12a}$$

(Visser, 1972).

Hamaker already indicated the possibility of such repulsive (dispersion) forces (1937) as did Derjaguin (1954); see van Oss *et al.* (1979) for experimental verification of this effect. All of these considerations initially only applied to van der Waals-London interactions, utilizing the Hamaker constant combining rules. The surface thermodynamic approach for obtaining Eqs. 5.10 and 5.11 is essentially the same as Hamaker's approach (1937), as long as Eq. 5.7 is valid. This validity however remains strictly limited to Lifshitz-van der Waals (LW) interactions.

5.2 The Lifshitz Approach

Lifshitz (1955) approached the problem of the van der Waals interaction by examining the macroscopic properties of materials, as opposed to the Hamaker treatment of summing individual atomic interactions. The derivation is based on Maxwell's equations modified to allow rapid temporal fluctuations (Rytov, 1959). This gives an approximate expression for the free energy of interaction between two different (1 and 2) semi-infinite surfaces separated by a third material (3) of thickness ℓ:

$$\Delta G_{132}(\ell) = \frac{kT}{\pi c^3} \int_{p=1}^{\infty} \sum_{n=0}^{\infty,} \varepsilon_3^{3/2} \omega^2 n \int_{\ell}^{\infty} P2 \left[\frac{\exp(2P\omega_n \ell \varepsilon_3^{1/2} e)}{\Delta_1 \Delta_2} - 1 \right]^{-1} dP d\ell \tag{5.13}$$

where k is Boltzmann's constant, T is the absolute temperature, c is the velocity of light, and P is an integration parameter. The values of Δ_1 and Δ_2 are:

$$\Delta_1 = \frac{\varepsilon_1(i\omega_n) - \varepsilon_3(i\omega_n)}{\varepsilon_1(i\omega_n) + \varepsilon_3(i\omega_n)} \tag{5.14}$$

$$\Delta_2 = \frac{\varepsilon_2(i\omega_n) - \varepsilon_3(i\omega_n)}{\varepsilon_2(i\omega_n) + \varepsilon_3(i\omega_n)} \tag{5.15}$$

$$\omega_n = \frac{4\pi^4 nkT}{h} \tag{5.16}$$

where h is Planck's constant, ω_n is the frequency, n the quantum number of the relevant oscillation and $\varepsilon(i\omega_n)$ the dielectric susceptibility along the complex frequency axis. The prime in the summation symbol of Eq. 5.13 indicates that the zero term is to be divided by two.

Integration of Eq. 5.13 yields:

$$\Delta G_{132}(\ell) = \frac{kT}{8\pi\ell^2} \sum_{n=0}^{\infty,} \sum_{\jmath=1}^{\infty} (\Delta_1\Delta_2) \left(\frac{X_o}{\jmath^2} + \frac{1}{\jmath^3}\right) \exp(-\jmath X_o) \tag{5.17}$$

where,

$$X_o = (2\omega_n \ell \varepsilon_3^{1/2})/c \tag{5.18}$$

and $\jmath = 1, 2, 3...$

As the separation distance, $\ell \to 0$, $X_o \to 0$ and we obtain a relation for the non-retarded van der Waals interaction energy:

$$\Delta G_{132}(\ell) = \frac{kT}{8\pi\ell} \sum_{n=0}^{\infty,} \sum_{\jmath=1}^{\infty} \frac{(\Delta_1\Delta_2)}{\jmath^3} \tag{5.19}$$

If the two interacting materials are the same (i.e., $1 = 2$) and they are interacting in a vacuum, Eq. 5.18 reduces to:

$$\Delta G_{11}(\ell) = \frac{kT}{8\pi\ell^2} \sum_{n=0}^{\infty,} \sum_{\jmath=1}^{\infty} \left[\frac{\varepsilon_1(i\omega_n) - 1}{\varepsilon_1(i\omega_n) + 1}\right]^{2\jmath} \jmath^{-3} \tag{5.20}$$

Combining Eq. 5.19 with the Hamaker constant (Eq. 5.8) gives:

$$A_{ii} = \frac{3kT}{2}\sum_{n=0}^{\infty,}\sum_{j=1}^{\infty}\left[\frac{\varepsilon_1(i\omega_n)-1}{\varepsilon_1(i\omega_n)-1}\right]^{2j} j^{-3} \tag{5.21}$$

where A_{ii} is the Hamaker constant for material i *in vacuo*. With the assumption that the ultraviolet part of the spectrum is the major contributor (Israelachvili, 1974), and considering only $j = 1$, Eq. 5.20 reduces to:

$$A_{ii} = \frac{3}{16\sqrt{2}}\frac{(n_o^2-1)^2}{(n_o^2+1)^{3/2}}n\omega_{UV} \tag{5.22}$$

With this expression and observed values of the index of refraction, assuming $\omega = 2.63 \times 10^{16}$ $rad\ sec^{-1}$, Israelachvili (1974) calculated the Hamaker constants for a number of liquids and also their surface tension from:

$$\gamma_i = \frac{A_{ii}}{24\pi\ell_o^2} \tag{5.23}$$

where γ_i is the apolar surface tension component for material i. In these calculations, the Born repulsion was not considered; the value of ℓ_o still was only roughly estimated, at about 2 Å; see also van Oss *et al.* (1979).

5.3 Interfacial Lifshitz-van der Waals Interactions

Following Lifshitz (1955), Chaudhury (1984) showed that the three types of van der Waals forces can be treated in the same manner so that they can be lumped together as a single term, the "apolar" term. van Oss *et al.* (1988a) demonstrated that ℓ_o has a value of 1.57 Å ± 0.09. Thus, the Hamaker constant of a given compound, i, can be directly obtained from its apolar (Lifshitz-van der Waals) surface tension component, γ_i^{LW}, cf. Eq. 5.23. The surface tension (γ_i), i.e., the surface free energy per unit area, of a liquid, *in vacuo*, is equal to one half the free energy of cohesion (ΔG_{ii}), and opposite in sign (Good, 1967).

$$\gamma_i = -\frac{1}{2}\Delta G_{ii} \tag{5.24}$$

For solids, Eq. 5.24 is equally true, but solids differ from liquids in that ΔG_{ii} is not their free energy of cohesion, but just the free energy available for interacting with liquids (Giese *et al.*, 1996). If the contributors to ΔG_{ii} are independent, as is commonly assumed, then it follows that the surface tension is composed of independent contributors each of which can be treated separately (Fowkes, 1963), i.e.:

$$\gamma_i = \sum_j \gamma_i^j \tag{5.25}$$

where the γ^j terms represent the specific contributors, e.g., hydrogen bonding, dipolar, dispersion, metallic, etc.

For purely Lifshitz-van der Waals interactions (symbolized by LW), the interfacial LW component of the surface tension for two materials, 1 and 2, can be obtained from the LW surface tension components of each by application of the Good-Girifalco-Fowkes combining rule (Good and Girifalco, 1960; Fowkes, 1963):

$$\gamma_{12}^{\mathrm{LW}} = \left(\sqrt{\gamma_1^{\mathrm{LW}}} - \sqrt{\gamma_2^{\mathrm{LW}}}\right)^2 \tag{5.26}$$

or:

$$\gamma_{12}^{\mathrm{LW}} = \gamma_1^{\mathrm{LW}} + \gamma_2^{\mathrm{LW}} - 2\sqrt{\gamma_1^{\mathrm{LW}}\gamma_2^{\mathrm{LW}}} \tag{5.27}$$

The apolar component of the free energy of cohesion of material 1 is:

$$\Delta G_{11}^{\mathrm{LW}} = -2\gamma_1^{\mathrm{LW}} \tag{5.28}$$

and the free energy of interaction between two materials is related to the surface tensions of these materials by the Dupré equation (Dupré, 1869):

$$\Delta G_{12}^{\mathrm{LW}} = \gamma_{12}^{\mathrm{LW}} - \gamma_1^{\mathrm{LW}} - \gamma_2^{\mathrm{LW}} \tag{5.29}$$

For two similar objects (1) immersed in a liquid (2), the relation is:

$$\Delta G_{121}^{\mathrm{LW}} = -2\gamma_{12}^{\mathrm{LW}} \tag{5.30}$$

and two different objects (1 and 2) immersed in a liquid (3) are related to the interfacial tensions by:

$$\Delta G_{132}^{\mathrm{LW}} = \gamma_{12}^{\mathrm{LW}} - \gamma_{13}^{\mathrm{LW}} - \gamma_{23}^{\mathrm{LW}} \tag{5.31}$$

Using Eqs. 5.26 and 5.29 to expand the interfacial surface tensions in Eq. 5.31 gives:

$$\Delta G_{132}^{\mathrm{LW}} = -2\gamma_3^{\mathrm{LW}} - 2\sqrt{\gamma_1^{\mathrm{LW}}\gamma_2^{\mathrm{LW}}} + 2\sqrt{\gamma_1^{\mathrm{LW}}\gamma_3^{\mathrm{LW}}} + 2\sqrt{\gamma_2^{\mathrm{LW}}\gamma_3^{\mathrm{LW}}} \tag{5.32}$$

Since:

$$\Delta G_{12}^{LW} = -2\sqrt{\gamma_1^{LW}\gamma_2^{LW}} \tag{5.33}$$

it follows from Eq. 5.32 that:

$$\Delta G_{132}^{LW} = \Delta G_{33}^{LW} + \Delta G_{12}^{LW} - \Delta G_{13}^{LW} - \Delta G_{23}^{LW} \tag{5.34}$$

which is the confirmation of the Hamaker combining rule (Eq. 5.11a) obtained via a purely surface thermodynamic treatment (Neumann *et al.*, 1982), based on 1) the applicability of the geometric mean combining rule to LW interactions and 2) the assumption that the value of ℓ_o is the same for all types of ΔG interactions (van Oss and Good, 1984). The constancy of ℓ_o for LW interactions of all materials was confirmed by van Oss *et al.* (1988a); see also van Oss (1994), Ch. XI.

5.4 Polar Forces

For some time it was thought that the Keesom dipole-dipole interactions should be treated separately from the Debye and London interactions. Because of the dipolar nature of the Keesom phenomenon, the term "polar" was applied to these interactions, in contrast to the "apolar" Debye and London interactions. This distinction between all three apolar electrodynamic forces impeded progress in the search for the true polar surface interactions. After Chaudhury (1984) showed that the three, apolar, electrodynamic forces are simply additive, and should be treated as a single entity, the LW interactions, it became possible to examine the nature of the polar (Lewis) properties of surfaces as an entirely separate phenomenon from their electrodynamic (LW) properties.

In aqueous media, and especially for solid surfaces which are rich in oxygen such as silicate minerals, the principal polar interaction is hydrogen bonding, involving donors and accepters. As this type of interaction can be treated as occurring between a Brønsted acid (the hydrogen donor) and a Brønsted base (the hydrogen acceptor), polar interactions must account for the dual nature of such interactions. Moreover, polar surface interactions are not restricted to hydrogen bonding, so that the polar concept has been extended to include all electron donating and electron accepting phenomena, as encompassed in the more general acid-base paradigm of Lewis (Fowkes, 1987; van Oss *et al.*, 1987a, 1988a; van Oss and Good, 1990). To emphasize the (Lewis) acid-base character of the polar interactions, the designation AB has been used.

The polar and apolar components of the free energies of interaction are additive (Fowkes, 1983; van Oss *et al.*, 1987a, 1988a,b):

$$\Delta G = \Delta G^{\text{LW}} + \Delta G^{\text{AB}} \tag{5.35}$$

and, rewriting Eq. 5.24 as

$$\Delta G_{\text{ii}} = -2\gamma_{\text{i}} \tag{5.36}$$

it follows that:

$$\gamma_{\text{i}} = \gamma_{\text{i}}^{\text{LW}} + \gamma_{\text{i}}^{\text{AB}}. \tag{5.37}$$

Because the electron donating and the electron accepting sites at a surface or interface are different and play different roles in the interfacial interactions, the polar properties of a surface are inherently asymmetrical and *must be described by two parameters.* This is very different from the LW interactions where, for example, (Eq. 5.33):

$$\Delta G_{\text{ij}}^{\text{LW}} = -2\sqrt{\gamma_{\text{i}}^{\text{LW}}\gamma_{\text{j}}^{\text{LW}}}. \tag{5.33a}$$

Such a simple combining rule is not applicable to AB interactions. For the AB interactions we define the free energy of interaction between two materials, *i* and *j* as:

$$\Delta G_{\text{ij}}^{\text{AB}} \equiv -2\sqrt{\gamma_{\text{i}}^{\ominus}\gamma_{\text{j}}^{\oplus}} - 2\sqrt{\gamma_{\text{j}}^{\ominus}\gamma_{\text{i}}^{\oplus}}. \tag{5.38}$$

where the electron donor parameter is designated as $\gamma^{\ominus}$ and the electron acceptor parameter is designated as $\gamma^{\oplus}$. The factor of 2 in the relation is introduced to maintain comparable magnitudes for $\gamma_{\text{i}}^{\ominus}$, $\gamma_{\text{i}}^{\oplus}$ and $\gamma_{\text{i}}^{\text{AB}}$ (van Oss *et al.*, 1987a). The polar component of the surface tension of compound, *i*, then is:

$$\gamma_{\text{i}}^{\text{AB}} \equiv 2\sqrt{\gamma^{\oplus}\gamma_{\text{i}}^{\ominus}}. \tag{5.39}$$

Note that Eq. 5.38 is very different from the LW component of the free energy of interaction (Eq. 5.33a). There are two terms in Eq. 5.38 because there exist two kinds of interactions each of which must be accounted for, i.e., an electron donor of *i* interacting with an electron acceptor of *j* (the first term) and an electron donor of *j* interacting with an electron acceptor of i (the second term). To attempt, as is sometimes done, to treat the polar interactions with a single term leads to profoundly erroneous results (van Oss and Good, 1992). Further, either term in Eq. 5.38 may be zero because, e.g. surface i may exhibit electron donating properties ($\gamma_i^{\ominus} > 0$) but no electron accepting properties ($\gamma_i^{\oplus} = 0$). Such a material is termed monopolar. For this situation:

$$\gamma_i^{AB} = 0 \tag{5.40}$$

and

$$\gamma_i = \gamma_i^{LW} \tag{5.41}$$

which is precisely what one would expect for a purely apolar surface. Thus, the condition in Eq. 5.40 should not be taken as an indication of the apolar nature of a material.

5.5 Lewis Acid-Base Interactions

The Dupré equation is equally applicable to polar interactions:

$$\Delta G_{ij}^{AB} = \gamma_{ij}^{AB} - \gamma_i^{AB} - \gamma_j^{AB} \tag{5.42}$$

so that the AB component of the interfacial tension can be written as:

$$\gamma_{ij}^{AB} = \Delta G_{ij}^{AB} + \gamma_i^{AB} + \gamma_j^{AB} \tag{5.43}$$

and, substituting the value for ΔG_{ij}^{AB} from Eq. 5.38 and the values for γ^{AB} from Eq. 5.39, gives:

$$\gamma_{ij}^{AB} = 2\left(\sqrt{\gamma_i^{\ominus}\gamma_i^{\oplus}} + \sqrt{\gamma_j^{\ominus}\gamma_j^{\oplus}} - \sqrt{\gamma_i^{\ominus}\gamma_j^{\oplus}} - \sqrt{\gamma_j^{\ominus}\gamma_i^{\oplus}}\right). \tag{5.44}$$

or,

$$\gamma_{ij}^{AB} = 2\left(\sqrt{\gamma_i^{\oplus}} - \sqrt{\gamma_j^{\oplus}}\right)\left(\sqrt{\gamma_i^{\ominus}} - \sqrt{\gamma_j^{\ominus}}\right). \tag{5.45}$$

Examination of the expression for the AB component of the interfacial tension (Eq. 5.45) shows that γ_{ij}^{AB} is not restricted to positive values or zero, as is the case for γ_{ij}^{LW}. Rather, γ_{ij}^{AB} will be negative when either:

$$\gamma_i^{\oplus} > \gamma_j^{\oplus} \text{ and } \gamma_i^{\ominus} < \gamma_j^{\ominus} \tag{5.46}$$

or:

$$\gamma_i^{\oplus} < \gamma_j^{\oplus} \text{and} \gamma_i^{\ominus} > \gamma_j^{\ominus}. \tag{5.47}$$

Since the AB and LW components of the interfacial tension are additive, the total expression for the interfacial tension between two condensed phases is:

$$\gamma_{ij} = \left(\sqrt{\gamma_i^{LW}} - \sqrt{\gamma_j^{LW}}\right)^2 + 2\left(\sqrt{\gamma_i^{\ominus}\gamma_i^{\oplus}} + \sqrt{\gamma_j^{\ominus}\gamma_j^{\oplus}} - \sqrt{\gamma_i^{\ominus}\gamma_j^{\oplus}} - \sqrt{\gamma_j^{\ominus}\gamma_i^{\oplus}}\right). \tag{5.48}$$

As will be shown in a subsequent chapter, the values of the surface tension components and parameters of an *unknown solid* are determined by measuring contact angles of liquid drops on the solid, and solving Young's equation (see Chapter 7). For this to work, the surface tension components and parameters of the test liquids must be known. This presupposes that one knows these values for a group of high energy liquids, i.e., those which will form finite contact angles on solid surfaces of interest. The values for the *liquids* can be obtained in a number of ways. For example, the total surface tension, γ, of the liquid can be derived from the shape of a pendant drop (Adamson, 1990). For apolar liquids, e.g., alkanes, this measurement gives the LW component, γ^{LW}, which completely characterizes the properties of the liquid. However, for polar liquids, the situation is more complex. Here, similarly, the total surface tension can be obtained directly: for water, as an example, $\gamma_w = 72.8$ mJ/m^2. The LW component can be obtained by contact angle measurements of the liquid on a low-energy, apolar solid such as Teflon. For water, $\gamma_w^{LW} = 21.8$ mJ/m^2 so that, by difference (Eq. 5.41), the AB component is found to be $\gamma_w^{AB} = 51$ mJ/m^2 (Fowkes, 1965; Chaudhury, 1984). However, there is no way to determine the values of both polar parameters from γ^{AB} as is seen from Eq. 5.39. It is convenient to set $\gamma_w^{\oplus} = \gamma_w^{\ominus} = 25.5$ mJ/m^2, at 20°C. While this choice is arbitrary, it has no effect on the values of, e.g., γ_{12}^{AB}, ΔG_{12}^{AB}, ΔG_{131}^{AB} and ΔG_{132}^{AB}, which are the values of interest (van Oss *et al.*, 1987a).

The $\gamma_L^{\oplus}$ and $\gamma_L^{\ominus}$ of polar liquids can be determined via contact angle determination on a known monopolar solid, such as polymethylmethacrylate (PMMA) which manifests only a (known) $\gamma_{PMMA}^{\ominus}$. From this measurement and the known γ_L^{LW} and γ_L^{AB} of the liquid, $\gamma_L^{\oplus}$ and $\gamma_L^{\ominus}$ can then be determined, using the Young Dupré equation (Eq. 6.4), see van Oss *et al.* (1990b).

5.6 Polar Attractions and Repulsion

It is of great importance that γ_{ij}^{AB} can be negative. The Dupré equation for three condensed media (of which at least one, i.e., #3, must be a liquid) gives us:

$$\Delta G_{132} = \gamma_{12} - \gamma_{13} - \gamma_{23}. \tag{5.49}$$

Expanding this in terms of the AB and LW components yields:

$$\begin{aligned}\Delta G_{132} = &\sqrt{\gamma_1^{LW}\gamma_3^{LW}} + \sqrt{\gamma_2^{LW}\gamma_3^{LW}} - \sqrt{\gamma_1^{LW}\gamma_2^{LW}} - \gamma_3^{LW} + \\ &2\sqrt{\gamma_3^{\oplus}}\left(\sqrt{\gamma_1^{\ominus}} + \sqrt{\gamma_2^{\ominus}} - \sqrt{\gamma_3^{\ominus}}\right) + \\ &2\sqrt{\gamma_3^{\ominus}}\left(\sqrt{\gamma_1^{\oplus}} + \sqrt{\gamma_2^{\oplus}} - \sqrt{\gamma_3^{\oplus}}\right) - \\ &2\sqrt{\gamma_1^{\oplus}\gamma_2^{\ominus}} - 2\sqrt{\gamma_1^{\ominus}\gamma_2^{\oplus}}\end{aligned} \tag{5.50}$$

Similarly, the interaction energy between two identical materials, 1, immersed in liquid 3 gives:

$$\begin{aligned}\Delta G_{131} &= -2\gamma_{13} \\ &= -2\left(\sqrt{\gamma_1^{LW}} - \sqrt{\gamma_3^{LW}}\right)^2 - \\ &4\left(\sqrt{\gamma_1^{\oplus}\gamma_1^{\ominus}} + \sqrt{\gamma_3^{\oplus}\gamma_3^{\ominus}} - \sqrt{\gamma_1^{\oplus}\gamma_3^{\ominus}} - \sqrt{\gamma_3^{\oplus}\gamma_1^{\ominus}}\right)\end{aligned} \tag{5.51}$$

and, lastly, the interaction between two different substances, 1 and 2, *in vacuo* is:

$$\Delta G_{12} = -2\left(\sqrt{\gamma_1^{LW}\gamma_3^{LW}} + \sqrt{\gamma_1^{\oplus}\gamma_2^{\ominus}} + \sqrt{\gamma_2^{\oplus}\gamma_1^{\ominus}}\right) \tag{5.52}$$

It is clear from Eq. 5.52 that the sign of the interaction energy between any two materials *in vacuo* is always negative, i.e., there is an attraction between them, and cannot be zero because γ^{LW} for all materials is finite and positive. The minimum energy one can have is roughly -40 mJ/m^2 (between, e.g., the apolar materials octane and Teflon) and also for water (a very polar material) interacting with Teflon, $\Delta G_{12} \approx -40\text{mJ/m}^2$. One normally considers Teflon to be a very "hydrophobic" material, yet it clearly has a substantial interaction energy with water. Hence the terms "hydrophobic" and "hydrophilic" are misleading as commonly used. In a later section, these two terms will be given a more reasonable definition.

5.7 Electrostatic Interactions

The above discussion concerning the interaction energy (ΔG_{132} and ΔG_{121}) between particles immersed in a liquid still is incomplete if these particles carry an electrostatic charge, in which case an additional term must be added which expresses the electrostatic contribution, termed EL, to the total interaction energy, expressed in general terms:

$$\Delta G = \Delta G^{LW} + \Delta G^{AB} + \Delta G^{EL} \tag{5.53}$$

5.8 Ionic Double Layer

The inclusion of the EL contribution is especially important when the particles are silicate minerals immersed in water. The origins of the electrostatic charge on such particles are diverse. They may arise from the dissociation of surface acidic or basic moieties, incongruous dissolution at the solid-water interface, a permanent structural imbalance in charge (e.g., the layer charge of phyllosilicate minerals), broken bonds at the interface, and selective adsorption of ions from the aqueous environment (Everett, 1988). Some or perhaps all of these factors contribute to the existence of a charge at the surface of the solid, which is the potential of the particle, ψ_o, and it is this quantity which is the basis for the interaction energy calculation. As a result of the existence of the electrostatic charge, the distribution of ions and water molecules in the vicinity of the particle-water interface is perturbed, and the perturbation dies off with increasing distance from the interface. This creates a diffuse double layer of ions and water molecules at and near the solid-liquid interface. For the simple situation of a uniform charge at the particle surface, the electrostatic potential decays exponentially with distance from the particle surface. The arrangement of ions and counterions whose distribution is different from the bulk liquid constitutes the diffuse double layer, arranged about the particle. A measure of the thickness of the double layer is given by the Debye length, $1/\kappa$, which is the distance from the surface to where ψ_o has decayed to a value of ψ_o/e.

The potential at the slipping plane is the ζ-potential, a quantity which can be determined by electrokinetic measurements. The relation between $\psi_o$and ζ for a spherical particle (for relatively small ζ-potentials) is:

$$\psi_o = \zeta \left(1 + \frac{z}{a}\right) e^{\kappa z} \tag{5.54}$$

where z is the distance from the surface to the slipping plane (usually 3 to 5 Å), and a is the radius of the particle; κ is the inverse thickness of the diffuse ionic double layer, or the Debye length. The Debye length $1/\kappa$ is expressed as:

$$\frac{1}{\kappa} = \left[(\varepsilon kT)/(4\pi e^2 \sum v_i^2 n_i^2)\right]^{1/2} \tag{5.55}$$

where ε is the dielectric constant for the liquid, k is Boltzmann's constant, T is the absolute temperature, e is the charge of the electron, and v_i is the valency of the ith ionic species with a concentration n_i (i.e., the number of ions per cm^3 of bulk liquid).

In aqueous systems, a problem arises in relating ψ_o and ζ in that the water lying within the slipping plane may be strongly oriented by interactions with surface atoms of the particle. The oriented structure of this water will reduce the value of the dielectric constant from the bulk value of 80. The difficulty is that there is no easy way to determine the value of ε within the slipping plane because this material is not readily accessible to experiment, thus introducing an uncertainty in the calculation of ψ_o. One solution to this is simply to treat the slipping plane as the surface of the particle and use ζ in place of ψ_o. This seems reasonable because the material within the slipping plane is really part of the particle within the context of a dilute aqueous suspension. When particles approach each other to distances which are less than twice the distance between the surface of the particle and the slipping plane, the situation becomes more complex. For a more thorough discussion of ψ_o and ζ-potentials, see Overbeek (1952), Shaw (1969), Sparnaay (1972) and Hunter (1981).

5.9 Electrokinetic Phenomena

Since a charged particle will be acted upon by an external electric field, the determination of the particle charge is ultimately based on the measurement of the movement of some object in response to the external field. The most straight-forward experiment is to observe the mobility of the particle itself in response to the field, i.e., *electrophoresis*. Alternately, the flow of the liquid in contact with a charged but immobile surface can be measured, by *electroosmosis*. If such a flow is blocked, the pressure build-up can be measured as the *electroosmotic counter-pressure*. Instead of allowing the polar fluid to flow past the charged surface, it can be forced to flow by pumping, in which case the resulting potential (*streaming potential*) is measured. Finally, the electric potential generated by the migration, most often by gravitational settling, of charged particles in a polar liquid is the *migration* (or *sedimentation*) *potential*.

5.10 The ζ-Potential

In treating electrophoretic mobility, there are a few assumptions made which are generally applicable to mineral particle suspensions. These are that the particles are rigid, non-conducting, their movement through the polar liquid is non-turbulent, there is either no electroosmotic back-flow along the walls of the vessel in which the measurement is made or the back-flow is accounted for, and the Brownian motion of the particles is unimportant.

Two cases are of importance for the discussion, based on the thickness of the double layer.

5.10.1 Thick double layer

For a thick (but insubstantial) double layer, there is no significant impediment to the electric field by the charged particles. The charged particle is acted upon by three forces. These forces include the attractive force between the particle and the oppositely charged electrode:

$$F_1 = ne\chi \tag{5.56}$$

where ne is the total effective electric charge on the particle and χ is the electric field strength. The second force is the hydrodynamic friction experienced by the particle:

$$F_2 = -6\pi\eta a U \tag{5.57}$$

where η is the viscosity of the fluid, a is the radius of the spherical particle (or some other expression of the dimension of a non-spherical particle) and

U is the velocity of the particle. Finally, there is the force of electrophoretic friction resulting from the movement of counterions through the fluid:

$$F_3 = \chi\,(\varepsilon\zeta a - ne) \tag{5.58}$$

where ε is the dielectric constant of the fluid and ζ the potential of the particle at the slipping plane. At equilibrium, these *three* forces should sum to zero (Huckel, 1924). Defining the electrophoretic mobility as:

$$u = \frac{U}{\chi} \tag{5.59}$$

and summing Eqs. 5.56, 5.57, and 5.58, one obtains:

$$u = \frac{\zeta\varepsilon}{6\pi\eta}. \tag{5.60}$$

Thus, under the conditions where the particle dimensions are small compared to the thickness of the double layer, the electrophoretic mobility, u, is directly proportional to the ζ-potential.

5.10.2 Thin double layer

In the case of a thin and dense double layer, the electric field is distorted by the particles in the fluid because of the (usually) larger particle size and the compactness of the double layer (von Smoluchowski, 1917, 1921). The principal result of this is that the electrophoretic friction, F_3, is considerably reduced because the oppositely charged ions in the path of the moving particle do not interfere with the particle's motion, and those ions slightly tangential to the particle contribute little to the friction. This increases the mobility of these particles. Here the electrophoretic mobility is derived using Poisson's equation:

$$\frac{d}{dx}\left(\varepsilon\frac{d\psi_o}{dx}\right) = -4\pi\sigma \tag{5.61}$$

where x is the distance from the slipping plane, ψ_o is the electric potential of the double layer, and σ is the surface charge. Integration for the forces acting on all volume elements and equating ζ with ψ_o yields (von Smoluchowski, 1917, 1921; Overbeek and Wiersema, 1967; Shaw, 1969):

$$u = \frac{\zeta\varepsilon}{4\pi\eta} \tag{5.62}$$

This is seen to differ from the Hückel equation (Eq. 5.60) only by a constant in the denominator. Henry (1931) resolved the two versions of the electrophoretic mobility relation by accounting for the radius of the particle and the thickness of the double layer, in terms of $1/\kappa$, as:

$$u = \frac{\zeta\varepsilon}{6\pi\eta} f(\kappa a) \tag{5.63}$$

where $f(\kappa a)$ varies between the limits of 1.5, for $\kappa a \to \infty$ (or, in practice, when $\kappa a > 300$), and 1.0, for $\kappa a \to 0$ (or, in practice, when $\kappa a < 1$). These limits correspond to von Smoluchowski's equation and to Hückel's equation, respectively. Thus, when the particle dimensions are large with respect to the thickness of the double layer, the electrophoretic mobility, u, is also proportional to the ζ-potential.

5.10.3 Relaxation

The preceding discussion presupposes that the double layer surrounding the charged particle, as it moves through the liquid, remains undistorted. However, such a distortion occurs, e.g., when the double layer moves at a slower velocity than the particle and thus is thinner at the leading edge. *Relaxation* effects then occur. The conditions under which relaxation can occur are described by Overbeek (1943) and Overbeek and Wiersema (1967) and involve both the values of the ζ-potential and the ratio κa. Relaxation is negligible for all values of the ζ-potential when κa is either small (< 0.1) or large (> 300), or when $\zeta < 25$ mV for all values of κa. That is, when the ζ-potential is small, the electrophoretic mobility is accordingly small and there is little distortion of the double layer about the moving particle. Similarly, there will be little distortion if the double layer is thick but insubstantial *or* thin and compact.

Otherwise, relaxation can seriously influence the electrophoretic mobility of a particle (Overbeek, 1943; Booth, 1948, 1950; Wiersema, 1964; Overbeek and Wiersema, 1967; Derjaguin and Dukhin, 1974) Solutions to this problem are reviewed by van Oss (1975).

5.11 Energy Balance Relationships

Thus far the discussion has dealt with the interfacial interactions between condensed media at contact. The net interfacial interaction is formulated as the value of ΔG, and the sign indicates whether there is a net attraction (negative sign) or a net repulsion (positive sign), according to the usual convention of chemical thermodynamics. To take an example, one could evaluate the attraction between two clay particles (1), in contact with each other, in the presence of a liquid such as water (2) by calculating $\Delta G_{121}^{\mathrm{Tot}}$. Suppose $\Delta G_{121}^{\mathrm{Tot}}$ for this system is negative, i.e., the clay particles adhere to each other when in contact. While interesting, this calculation is only part of the story; there is no guarantee that when these particles are separated by a given distance, say 50 Å, there will be an attraction. In short, we must examine how the free energy of the system varies as the distance between particle pairs changes. This variation can be examined in the form of an energy balance plot of free energy versus interparticle separation.

5.12 Decay with Distance

The value of the free energy as a function of distance is determined by two factors; the total interaction energy between the particles while at contact and the manner by which this energy decays with increasing distance. The form of the decay function depends on the shape and size of the particle and the type of interaction (e.g., LW). While one can accurately determine ΔG^{Tot} and its components for clay particles in contact, the manner in which the interaction energy decays with distance is only an approximation because the decay formulae describe only simple geometrical shapes whilst clay particles rarely assume these shapes. However, the situation may not be entirely hopeless as will be discussed later.

5.12.1 LW interactions

Neutral, spinless atoms have a zero dipole moment, but this is, in quantum mechanical terms, only an average value. Spontaneous dipole moments can and do arise continually and these can interact with neighboring atoms or molecules to induce a dipole moment such that they can interact and attract each other. This interaction between objects is propagated at the speed of light so that, for relatively small interparticle distances, the two objects will be correlated and a van der Waals-London attraction results. As the distance between the particles increases, the length of time for interaction also increases and the correlation decreases. Thus, there are two regimes; the non-retarded regime where the distances are short enough for correlation to exist, and the retarded regime where the correlation is decreased. The boundary between these two is of the order of 50 to 100 Å. However, one is normally interested in the interaction between particles at distances shorter than 100 Å because this is usually the region which is critical for determining whether a system of particles is stable or unstable.

It is convenient to combine Eqs. 5.23 and 5.26 to give the surface free energy in terms of the Hamaker constant.

$$\Delta G_{ii}^{LW} = -2\gamma_{i}^{LW} = -\frac{A_{ii}}{12\pi\ell_o^2} \tag{5.64}$$

The value of ℓ_o has been estimated from a comparison of A_{ii} and ΔG_{ii}^{LW} values for a wide range of materials (e.g., helium, hexane, glycerol) giving a value of 1.57 ± 0.09 Å. Thus, the value of ΔG_{ii}^{LW} in Eq. 5.64 corresponds to the LW component of the cohesion energy between two objects in contact (i.e., at a distance $\ell = \ell_o = 1.57$ Å). How this value decays with distance can be determined using the method of Hamaker (1937) (see Nir, (1976)). For example, the interaction between two semi-infinite (i.e., thick) plane parallel slabs as a function of separation is given by:

$$\Delta G_{\ell}^{LW} = \Delta G_{\ell_o}^{LW}\left(\frac{\ell_o}{\ell}\right)^2 = -\left(\frac{A}{12\pi\ell^2}\right). \tag{5.65}$$

5.12.2 Polar interactions

The mechanism by which particles interact in a polar sense is very different from that of the apolar (electrodynamic) interactions. In the case of polar interactions, one needs the intervention of a continuous polar medium; typically the very polar fluid, water. A polar interaction at the interface between a particle and an adjacent water molecule causes the latter to orient, creating an ordered (i.e., structured) layer of water. If the interfacial interaction is sufficiently strong, the first layer of almost completely ordered water will, in turn, create order in the next layer, and so on. At each successive layer of water, the magnitude of the interaction decreases, resulting in a decay. The scale over which this interaction will operate is given by the correlation length, λ for the liquid medium. Pure water has a value of $\lambda \approx 2$ Å according to Chan *et al.* (1979) but when hydrogen bonding between water molecules is taken into account, λ is closer to 6 to 10 Å. For the situation where the solid surface is moderately hydrophobic or hydrophilic, a value of $\lambda \approx 6$ to 10 Å thus seems more reasonable. In practice, it is usually best to assume $\lambda = 10$ Å. The assumption of much larger values for λ larger than 10 Å in

Particle Geometry	$\Delta G_{\ell}^{LW} =$	$\Delta G_{\ell}^{AB} =$	$\Delta G_{\ell}^{EL} =$
parallel, flat plates	$-A/12\pi\ell^2$	$\Delta G_{\ell_o}^{AB''} X$	$\frac{64}{\kappa}\mathrm{nkT}\gamma_o^2 \exp(-\kappa\ell)$
sphere with flat plate or cylinders crossed at 90°	$-AR/6\ell$	$2\pi R\lambda\Delta G_{\ell_o}^{AB''} X$	$\varepsilon R\psi_o^2 Y$
sphere with sphere	$-AR/12\ell$	$\pi R\lambda\Delta G_{\ell_o}^{AB''} X$	$\frac{1}{2}\varepsilon R\psi_o^2 Y$

n = the number of counterions per cm^3

A = Hamaker constant, defined in Eq. 5.64

k = Boltzman's constant

T = temperature in K

$\gamma_o^2 = [\exp(ve\psi_o/2\mathrm{kT}) - 1]/[\exp(ve\psi_o/2\mathrm{kT}) + 1]$

where v = valence of the counterions

and e = charge of the electron

$X = \exp[(\ell_o - \ell)/\lambda]$

$Y = \ln[1 + \exp(-\kappa\ell)]$

ϵ = dielectric constant of the liquid medium; for water $\epsilon = 80$, at 20°C

Table 5.1: *Decay relations for the non-retarded LW interactions and for the AB and EL interactions with different particle geometries. For the sphere and cylinder geometries, R is the radius. It should be noted that all three classes of free energies involving spheres with radii, R, are proportional to R. The superscript AB'' indicates that the value of ΔG is measured at contact ($\ell = \ell_o$) in the flat parallel slab geometry.*

the case of hydrophobic interactions (see, e.g., Christenson (1992)) appear largely based upon experimental artefacts (Wood and Sharma, 1995), so that even for hydrophobic interactions $\lambda \approx 10$ Å should be maintained.

The decay of the AB interactions with distance is taken to be an exponential varying inversely with λ. For the same group of geometries, the decay formulae are listed in Table 5.1.

5.13 Electrostatic Interactions

As indicated in an earlier section, electrokinetic experiments measure the particle mobility when immersed in a liquid in response to an applied electric potential. From the mobility, the ζ-potential is calculated and the ζ-potential then allows calculation of the particle potential, ψ_o (cf. Eq. 5.54), which is

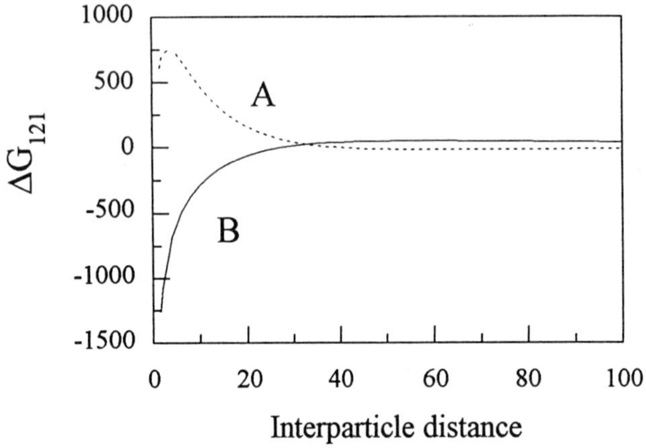

Figure 5.1: *Two interaction energy (in kT) curves as a function of interparticle distance (in Å) for a hydrophobic material (curve A) and a hydrophilic material (curve B).*

the fundamental quantity for calculating the interactions between particles. The electrostatic interaction energy decays as an exponential function of the interparticle distance, ℓ, and the inverse of the Debye length, κ. The decay relations are also listed in Table 5.1.

5.14 Energy Balance Diagrams

Given a simple situation of two particles immersed in a liquid, Eq. 5.53 describes how the individual contributions to the total interaction energy are combined. This relation is valid even if the two particles in question are not in contact. Thus, for each of the ΔG terms in Eq. 5.53, a term for ΔG_{ℓ} can be chosen from Table 5.1, as the geometry dictates. This allows the calculation of the total interaction energy for a given pair of particles in a liquid as a function of the interparticle distance. The form of the resulting function can be used to determine the stability (i.e., the particles will not be attracted to each other and will stay in suspension, neglecting gravitational forces) or instability (i.e., agglomeration or flocculation is favored). It is, incidentally, a fairly simple matter to distinguish between agglomeration or flocculation on

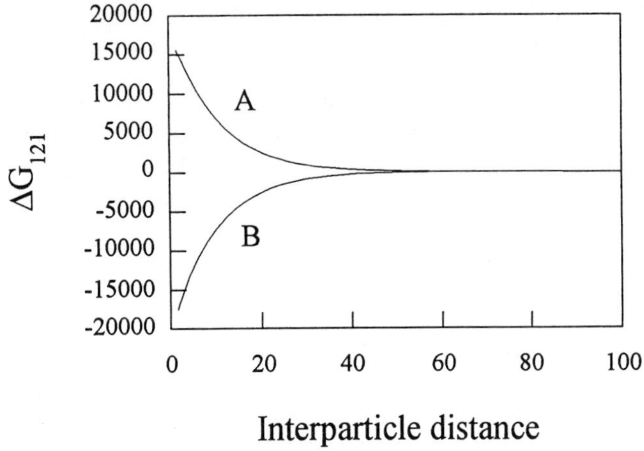

Figure 5.2: *Curve A is the case where there is a net repulsion at short interparticle distances but a minimum of attraction at larger distances. Curve B shows the reverse, a net adhesion between the particles at contact but close approach is prevented by an energy barrier.*

the one hand and sedimentation on the other. Agglomerated or flocculated complexes form a (microscopically visible) coarse, open structure, whereas gravitationally sedimented particles present as a densely packed mass.

5.14.1 Types of energy balance diagram

Most of the research in this area has involved aqueous electrolyte suspensions of colloidal-sized solid particles. Relatively less is known about the interaction of colloids when immersed in an organic liquid, and this is largely due to the paucity of measurements of the electrokinetic behavior of colloids in these liquids. Because the three types of interactions decay differently (see Table 5.1), the total interaction energy may exhibit inflections as the interparticle distance changes. One may usefully distinguish three types of energy balance diagrams. At one extreme, one would have the situation where there exists a large repulsion energy at contact (i.e., $\Delta G > 0$; $\ell = \ell_o$), sufficiently large so that the total interaction energy is always repulsive regardless of the interparticle distance. The other extreme occurs when the particles attract each other strongly enough so that there is a net attraction regardless of the interparticle distance. These two cases are shown in Figure 5.1.

Between these two extreme situations, one might expect to find a situation where the sign of the total interaction energy changes as ℓ increases. Figure 5.2 shows the two types of energy balance diagrams that are possible. In one case, curve A, there is a net attraction at contact, but this changes to a net repulsion at greater interparticle distances. Even though one might assume from the negative sign of ΔG that the particles would agglomerate, the positive values of ΔG at values of $\ell > \ell_o$, if sufficiently large, would prevent the particles from approaching each other. The existence of this energy barrier would stabilize the particles in the liquid at least temporarily. (This situation often occurs in aqueous suspensions of clay particles.)

The alternate possibility also exists, that of a positive value for ΔG at contact, changing to negative values as the interparticle distance increases. This would indicate a tendency for the particles to approach each other until the minimum of the interaction energy curve is reached, suggesting the formation of a stable arrangement of particles which are not in physical contact. (This is often the situation for suspended red blood cells and other platey particles).

There are uncertainties in the calculation of the energy balance diagrams, partly due to the approximate nature of the decay formulae of Table 5.1 and also to the fact that one rarely deals with particles whose shapes exactly match the geometries for which there are decay relations.

6

Measurement of Surface Thermodynamic Properties

The atoms which are at the external surface of a liquid or solid are in a very different environment compared to those atoms buried inside the object. This difference arises from the asymmetrical environment; in the bulk material, each atom is surrounded by similar atoms while those at the surface see this only on one side of the interface. In addition, the various influencing factors exerted by the environment act only on the outermost atoms. These atoms, as a consequence, have a different distribution from the inside, causing a different energy at the surface which is what one wants to measure. This energy is a function of the surface tension, γ, and its constituents, γ^{LW}, $\gamma^{\oplus}$ and $\gamma^{\ominus}$.

There is now a wealth of spectroscopic and other analytical techniques for probing the surface properties of solid materials (Vickerman, 1997; Riviere and Myhra, 1998), which yield a variety of surface properties of those parts of such materials that are situated anywhere between 1.0 and more than 10 nm *below their surfaces.* However, to date only contact angle analysis is capable of yielding the actual surface or interfacial properties *at the precise surface of solids*, that are germane to their interaction with other condensed-phase materials. Thus, methods pertaining to contact angle measurement and related techniques are the main emphasis of this chapter.

The following chapter deals with measurement methods of the electrokinetic surface potential, or ζ-potential of surfaces or interfaces of condensed phase materials.

6.1 The Young Equation

Contact angle (ϑ) measurement, first described by Thomas Young in 1805, remains at present the most accurate method for determining the interaction energy between a liquid (L) and a solid (S), at the minimum equilibrium distance between the liquid and the solid at the interface between the two. Young's paper contained no mathematical equations whatsoever, but its text expresses Young's proposition as follows:

$$\gamma_L \cos\vartheta = \gamma_S - \gamma_{SL} \tag{6.1}$$

where γ_{SL} represents the interfacial tension between the liquid and the solid. (Since the apparition in 1937 of a publication by Bangham and Razouk, who invoked the existence of a *spreading pressure*, π_e, caused by the evaporation of molecules of the liquid, L, emanating from the drop and re-condensing on the solid surface, S, many authors have added the subscript, V, to γ_L and γ_S, expressing them as γ_{LV} and γ_{SV}. As will be discussed below (section 6.4.2), this is an unnecessary complication which leads to confusion and is best avoided.

6.2 The Young-Dupré Equation as a Force Balance

In Eq. 6.1, γ_L (usually) and $\cos\vartheta$ are known and γ_S and γ_{SL} are the unknowns. Using two different liquids gives rise to two equations with three unknowns, i.e., γ_S, γ_{SL_1}, and γ_{SL_2}, where the subscripts 1 and 2 refer to the two equations (i.e., the two different liquids). Thus, Eq. 6.1, in the form given above, is not practically usable.

However, in conjunction with the Dupré equation in the following form:

$$\Delta G_{SL} = \gamma_{SL} - \gamma_S - \gamma_L \tag{6.2}$$

(see also Eqs. 5.29 and 5.42), Eq. 6.1 becomes:

$$(1 + \cos\vartheta)\gamma_L = -\Delta G_{SL} \tag{6.3}$$

which is known as the Young-Dupré equation. Eq. 6.3 has only one unknown; ΔG_{SL}. Combining Eqs. 6.1 and 6.2 with Eq. 6.3 one obtains:

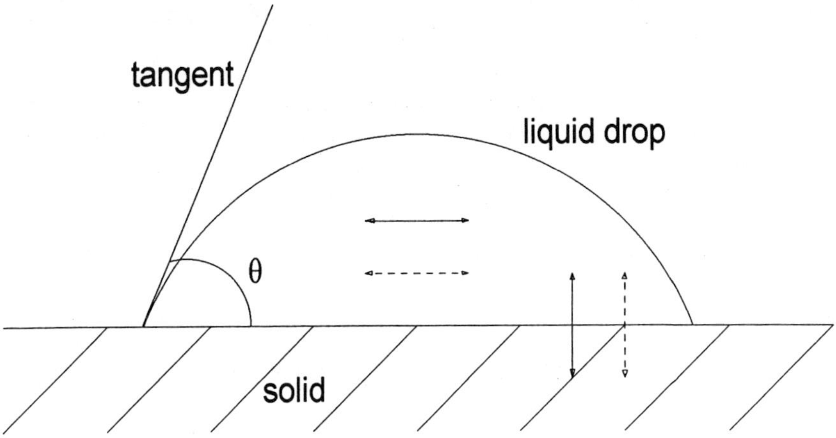

Figure 6.1: *The contact angle , ϑ, as a force balance. Cos ϑ is a measure of the equilibrium between the energy of cohesion between the molecules of liquid L (horizontal arrows) and the energy of adhesion (vertical arrows) between liquid L and solid S. The solid lines represent apolar interactions and dashed lines are polar interactions.*

$$\frac{1+\cos\vartheta}{2}\gamma_L = \sqrt{\gamma_S^{LW}\gamma_L^{LW}} + \sqrt{\gamma_S^{\oplus}\gamma_L^{\ominus}} + \sqrt{\gamma_L^{\oplus}\gamma_S^{\ominus}} \tag{6.4}$$

which is the complete Young-Dupré equation, applicable to apolar as well as to polar systems (van Oss *et al.*, 1987a, 1988a; van Oss 1994). Eq. 6.4, where the left hand side (containing the contact angle ϑ, measured through the drop, at the tangent to the drop, starting at the triple point, solid-liquid-air) represents the part of the energy of cohesion of the liquid, L, which is in equilibrium with and therefore equal to, the energy of adhesion between liquid, L, and solid, S, expressed in the right side of Eq. 6.4; see Figure 6.1.

Thus, the measured contact angle between L and S permits the use of the liquid-solid interaction as a force balance (cf. de Gennes (1990) and van Oss (1994)). It should be noted that Eq. 6.4 contains three unknowns, i.e., γ_S^{LW}, $\gamma_S^{\oplus}$, and $\gamma_S^{\ominus}$ (assuming that the values for γ_L^{LW}, $\gamma_L^{\oplus}$, and $\gamma_{SL}^{\ominus}$ are known as they would be in a real experiment). To obtain the three γ-component values for the solid, S, it is therefore necessary to measure contact angles with three *different* liquids, L_1, L_2, and L_3, of which at least two must be polar. A list of the most used contact angle liquids, with their γ-properties at 20°C, is given in Table 6.1. It should be emphasized that, to obtain finite, measurable contact angles γ_L must be greater than γ_S. When $\gamma_L < \gamma_S$, the liquid forms no contact angle on the solid, but spreads and wets it completely. The array of usable high-energy contact angle liquids is therefore quite limited. (The alkanes listed in Table 6.1 are not used normally for contact angle measurements, but for their wetting properties, used in thin layer wicking, described below).

6.3 Concept of the Surface Tension of a Solid

Whilst the surface tension of a liquid is correctly defined as:

$$\gamma_L = -\frac{1}{2}\Delta G_{LL}^{Cohesion} \tag{6.5}$$

(see Eq. 6.24), such a statement is only applicable to solids for Lifshitz-van der Waals interactions:

$$\gamma_S = -\frac{1}{2}\Delta G_{SS}^{CohesionLW} \tag{6.6}$$

For Lewis acid-base (or for that matter for covalent) interactions, the cohesive energy holding the solid together is not measurable by the contact angle approach. By contact angle measurements with polar liquids one determines the *excess* $\gamma_S^{\oplus}$ or, in most cases, the *excess* $\gamma_S^{\ominus}$ value of a *dry surface.* When, as is generally the case, in the polar part of the cohesion of a solid, when all the electron acceptors ($\gamma^{\oplus}$) are bound to an equivalent number of electron donors ($\gamma^{\ominus}$), there tend to be a sizable number of unbound electron-donors ($\gamma^{\ominus}$) left over (van Oss *et al.*, 1997). This excess $\gamma_S^{\ominus}$ is the only polar entity that can be measured via the contact angle approach, on dry solid surfaces. It is also the only $\gamma_S^{\ominus}$ by which the solid can interact with other polar entities, solid or liquid.

Thus, for any solid, γ_S is best defined (if one wishes to define it at all) as the value of $\gamma_S^{LW} + 2\sqrt{\gamma_S^{\oplus}\gamma_S^{\ominus}}$ that is obtained by contact angle determination. As a *dry solid* generally only manifests either a $\gamma_S^{\oplus}$ or $\gamma_S^{\ominus}$, its $\gamma_S = \gamma_S^{LW}$. When both $\gamma_S^{\oplus}$- and $\gamma_S^{\ominus}$-values are found for a solid surface, it is *wet*, usually with water. It is only in polar *liquids* that electron-acceptors ($\gamma_L^{\oplus}$) *and* electron-donors ($\gamma_L^{\ominus}$) can co-exist (van Oss *et al.*, 1997).

Liquid	γ	γ^{LW}	γ^{AB}	$\gamma^{\oplus}$	$\gamma^{\ominus}$	η
Apolar						
hexane	18.4	18.4	0	0	0	0.00326
octane	21.62	21.62	0	0	0	0.00542
decane	23.83	23.83	0	0	0	0.00907
dodecane	25.35	25.35	0	0	0	0.01493
tetradecane	26.6	26.6	0	0	0	0.02180
pentadecane	27.07	27.07	0	0	0	0.02841
hexadecane	27.47	27.47	0	0	0	0.03451
cis-decalin	32.2	32.2	0	0	0	0.03381
α-bromonaphthalene	44.4	44.4	0	0	0	0.0489
diiodomethane	50.8	50.8	0	0	0	0.028
Polar						
water	72.8	21.8	51.0	25.5	25.5	0.010
glycerol	64	34	30	3.92	57.4	14.90
formamide	58	39	19	2.28	39.6	0.0455
ethylene glycol	48	29	19	1.92	47.0	0.199

Table 6.1: *Surface tension components and parameters of liquids useful for contact angle and wicking measurements, at 20° C, in mJ/m^2, and their viscosities, in poises.*

6.4 Contact Angle Measurements on Heterogeneous Surfaces

6.4.1 The Cassie equation

On solid surfaces which are mosaics of different materials, e.g., an apolar material, 1, and a polar material, 2, Cassie (1948) showed that:

$$\cos\vartheta_A = \phi\cos\vartheta_1 + (1-\phi)\cos\vartheta_2 \tag{6.7}$$

where ϑ_A denotes the aggregate contact angle measured on the mosaic surface, and ϕ is the proportion of the surface occupied by material 1, while ϑ_1 and ϑ_2 are the contact angles one would find on a solid surface solely consisting of material, 1, and of material, 2, respectively.

As an example, a moderately hydrophobic, negatively charged clay particle, from Wyoming (SWy-1 from the Clay Minerals Repository) can be made much more hydrophobic by coating it with hexadecyl trimethyl ammonium cations (HDTMA) (Giese *et al.*, 1996). In this manner a proportion, ϕ, of the SWy-1 surface (largely the 001 surfaces) is coated with HDTMA. From the surface properties of untreated SWy-1 and of HDTMA-treated SWy-1 (Giese *et al.*, 1996) and from the surface properties of the hydrophobic tail of HDTMA alone, in the guise of hexadecane ($\gamma^{\mathrm{LW}} = 27.5$ mJ/m^2; $\gamma^{\oplus} = \gamma^{\ominus} = 0$), using Eq. 6.6, it can be calculated that approximately 39% of the total surface area of the SWy-1 was hydrophobized by means of electrostatically bound hexadecyl groups.

6.4.2 The extent to which solid surfaces become heterogeneous by condensation of molecules evaporating from the liquid drop

One suspected source of heterogeneity that could occur with even the most homogeneous surface, is the heterogeneity caused by the deposition of molecules emanating from the contact angle liquid, via the gaseous phase, by re-condensation. One school of thought, initiated by Bangham and Razouk (1937), has it that this condensation of liquid from the contact angle liquid gives rise to complete wetting of the solid surface (cf. Figure 6.1), which seriously reduces the contact angle. In view of this wetting effect, Bangham and Razouk introduced the terms γ_{LV} and γ_{SV} (for liquid-vapor and solid-vapor), instead of γ_{L} and γ_{S}. This nomenclature has been followed by many workers in the field of colloid and surface science, up to the present day.

It is easily shown however that no such drastic contamination by the contact angle liquid occurs in real life. From:

$$S + L \rightleftarrows SL \tag{6.8}$$

which describes the (reversible) adsorption reaction of the liquid onto the solid, the adsorption constant, K_a, can be expressed in relation to the free energy of adsorption as:

$$K_a = \exp\left(\frac{-\Delta G_{\mathrm{SL}}}{kT}\right) \tag{6.9}$$

where K_a is expressed as (mole fractions)$^{-1}$. K_a can also be expressed as:

$$K_a = \frac{\phi}{A^*(1-\phi)} \tag{6.10}$$

where ϕ (as defined before) is the proportion of a unit surface area of the solid, S, that it occupied by molecules of the vapor from liquid, L, and A* is

Liquid molecules with PMMA	ϕ (%)	ϑ_A (ϑ as measured)	ϑ_1^a ϑ for $\phi = 0$	$\Delta\vartheta$ ($= \vartheta_1 - \vartheta_A$)
diiodomethane	1.53	38.0°	38.3°	0.30°
1-bromonaphthalene	2.14	24.15°	24.42°	0.27°
water	3.55	72.67°	74.22°	1.55°
glycerol	0.062	67.24°	67.26°	0.02°
formamide	0.075	56.46°	56.48°	0.02°
ethylene glycol	2.30	76.72°	77.78°	1.07°
H_2O on PEO	3.90	18.02°	18.39°	0.37°

Table 6.2: *The percent coverage (ϕ) of polymethylmethacrylate (PMMA) with gas molecules emanating from various contact angle liquids, at 20°C and 76 cm Hg ambient atmospheric pressure, and the decrease in contact angle ($\Delta\vartheta$) that this coverage would give rise to.*

the ratio of molecules of the gas emanating from the liquid, to the number of air molecules, at ambient atmospheric pressure; A* follows from the vapor pressure for each given liquid. ΔG_{SL} is determined by contact angle measurement, using the Young-Dupré equation (Eq. 6.4), and ϕ then follows by combining Eqs. 7.12 and 7.13. Table 6.2 shows the values obtained for ϕ (in %) for the interaction of the contact angle liquids listed in Table 6.1, with the moderately hydrophobic polymer surface of poly(methyl methacrylate), PMMA, as well as for the interaction of water with the very hydrophobic polymer poly(ethylene oxide), PEO (van Oss *et al.*, 1998). In addition, Table 6.2 lists the difference in contact angle ($\Delta\vartheta$) caused by the deposition of liquid molecules from the gaseous state onto the solid surface. $\Delta\vartheta$ is found by inserting the value ϕ, for each case, in the Cassie equation (Eq. 6.7), where:

$$\Delta\vartheta = \vartheta_1 - \vartheta_2 \tag{6.11}$$

As shown in Table 6.1, the maximum percentage of contamination (ϕ) of surface, S, (in this case PMMA), by vapors from various contact angle liquids is at most only a few percent and in the cases of glycerol and formamide less than 0.1%. Even in the case of the most hydrophilic material known (PEO) the maximum amount of surface occupied by water is less than 4%. Expressed in decrease ($\Delta\vartheta$) in contact angle (ϑ as compared to the ideal contact angle (ϑ_1), this usually amounts to less than 1°, except in the case of water on PMMA, where it is 1.55°. It should be noted that the precision of contact angle measurement is rarely better than ± 1°. In virtually all cases the contamination of the surface by gas molecules emanating from the liquid drop can therefore be safely neglected. In the rare cases where ϑ has to be measured with an accuracy that is much better than $\pm$ 1° (e.g., for water on intermediate hydrophobic materials such as PMMA), the above treatment shows the methodology for calculating the value of a correction factor, based on the measured contact angles van Oss *et al.* (1998). It is also clear that the designations "γ_{LV}" and "γ_{SV}" instead of the simpler γ_L and γ_S are superfluous and misleading and thus are best avoided.

6.5 Contact Angle Measurement on Solid, Flat Surfaces

6.5.1 Advancing and retreating contact angles

The Young equation is held to be valid for contact angles measured as the *advancing* angle, i.e., the angle the drop makes when it has just ceased across the solid surface (e.g., for at most a few seconds; (Chaudhury, 1984)). Whilst advancing contact angles (ϑ_A) have been held to be a measure of the apolar aspect of a surface, and receding contact angles (ϑ_R) a measure of the polar aspects of the solid (Fowkes *et al.*, 1980), it would be hazardous to take that statement so literally as to attempt to derive the polar surface tension component of the solid by measuring ϑ_r. The difference between ϑ_a and ϑ_r is called the contact angle hysteresis. Another important cause for hysteresis is surface roughness. Methods have been worked out for applying correction factors to the measured (advancing) contact angles, once the radius of roughness has been determined (Chaudhury, 1984). However, using smooth surfaces, i.e., surface with radii of roughness significantly smaller than 1 μm, is by far preferable (Chaudhury, 1984). Various other possible causes for hysteresis are discussed by Adamson (1990).

Usually hysteresis, when it occurs, is positive, i.e., $\vartheta_a > \vartheta_r$, indicative of complete or partial residual wetting of the solid surface by the liquid of the retreating drop. However, negative hysteresis (i.e., $\vartheta_a < \vartheta_r$), although rare, can take place. For instance, when one measures the contact angle with diiodomethane on a fresh mica surface, cleaved only a second or so before the measurement, one finds a very small contact angle (indicating a γ^{LW}-value for the mica of 50 mJ/m^2 or more), after which ϑ very quickly increases to about 40°, corresponding to a γ^{LW}-value of the order of 40 mJ/m^2 (Giese *et al.*, 1996). This happens as a consequence of the rapid uptake of water of hydration (due to normal ambient humidity), once a mica surface is cleaved.

The much more common occurrence of positive contact angle hysteresis with polar liquids (e.g., with water) on polar surfaces is caused by the residual wetting the retreating liquid leaves behind. Thus, *only the advancing contact angle* (ϑ_a) *has significance* when used to obtain γ-values by means of the Young-Dupré equation (Eq. 6.4), because only an *advancing drop* encounters a new uncontaminated surface. The angle made by a *receding drop has no quantitative significance* in conjunction with Eq. 6.4, and may only offer an exceedingly rough qualitative indication of surface roughness, heterogeneity or partial hydrophilicity.

6.5.2 Preparation of solid surfaces

The easiest surface to prepare for contact angle measurements are smooth, flat surfaces of solid materials. These usually only need thorough cleaning, e.g., with chromic acid solution, followed by rinsing with distilled water, for glass, or treatment with acetone as applied to Teflon (Chaudhury, 1984). Under certain conditions, other types of surfaces may need even less preparation, for instance, surfaces on the inside of polystyrene Petri dishes which are hermetically sealed in sterile packs can be used without further preparation. Also, freshly cleaved mica may be presumed to be clean, but to avoid almost immediate contamination of such high-energy surfaces by various air pollutants, mica surfaces may have to be kept, and measured, in a vacuum. Lastly, for many purposes, it may be desirable to utilize the surface without preparation, as in the case of minerals where one wants to know the surface properties of the material as it exists in nature, i.e., in the normally contaminated condition.

For some solids (e.g., magnesium stearate and polyethylene oxide), available in powdered form, melting and casting smooth films by pouring the molten material onto a clean glass surface is a viable procedure. Powders such as the swelling (smectite) clays, or bacterial or mammalian cells, which can form reasonably confluent layers, are best deposited in a flat layer on a porous silver membrane by suction applied to an aqueous suspension of the particles. Subsequently, some air drying is required, until a constant contact angle with water is obtained (van Oss *et al.*, 1975; van Oss 1994).

Most micromolecular and macromolecular solutes can be transformed into smooth flat solid surface by deposition of a solution of such a solute in an appropriate solvent onto a glass surface, followed by evaporation of the solvent (in air or in *vacuo*). In most cases it is advisable to keep the dried surfaces (supported on the glass slides) in a vacuum desiccator. It usually is possible, upon drying, to obtain a very smooth, glassy surface with most polymers (e.g., water-soluble proteins and polysaccharides). However, with some organic polymers a fairly rough surface ensues upon drying from solution in a number of solvents. It is usually possible, however, to find a solvent from which a given polymer will dry as a very smooth surface. For instance, poly(methyl methacrylate) will yield rough surfaces when prepared from solutions in methyl ethyl ketone or chlorobenzene, but it will give a very smooth glassy surface when dried from a solution in tetrahydrofuran or in toluene.

To obtain flat layers of a *hydrated* biopolymer, membrane ultrafiltration should be used (in analogy with the deposition of layers of cells on microporous membranes). To that effect, a concentrated solution of, e.g., a given protein is ultrafiltered on an anisotropic membrane, which is impermeable to the protein, under a few atmospheres' pressure, until no further ultrafiltrate ensues (van Oss *et al.*, 1975; van Oss 1994). Here also, after a certain initial drying time, a plateau value persists for the contact angle for 30 minutes to one hour. The contact angle at the plateau value is consistent for any given protein, whereas plateau contact angles vary considerably among different proteins (van Oss *et al.*, 1975; van Oss 1994). By careful monitoring of the amount of residual water still present in the final retentate in the ultrafilter it is possible to determine the thickness of the hydration layer per protein molecule, e.g., in numbers of layers of water molecules of hydration (van Oss 1994).

Some smectite clay minerals are capable of forming self-supporting films by deposition of an aqueous suspension onto thin plastic films such as those marketed for wrapping food before refrigeration. If that fails, films of these minerals can be formed by suction through a silver membrane or through a Millipore filter of appropriate pore size (typically on the order of 0.2 μm). Alternatively, an aqueous suspension of the fine mineral particles can be transferred to clean glass microscope slides with a pipette and slow evaporation of the water will leave a uniform thin layer of the mineral powder. The only difficulty encountered in this process is to discover the concentration of powder in the water which gives the desired thickness and uniformity of film. These slides can be used either for direct contact angle measurements, or if they are too porous, thin layer wicking can be utilized to give the contact angles indirectly (see van Oss *et al.*(1992a) and Li *et al.*(1994)).

In principle, it is possible to load fine powders into narrow glass capillaries and to measure the rate of capillary rise, giving the contact angle via the Washburn equation. In practice, the packing of the powder particles is rarely dense enough and what usually occurs is that the passage of the liquid through the powder column results in a rearrangement of the particles to a closer packing, resulting in a rupture in the continuity of the powder and the cessation of the movement of the liquid. However, with spherical monosized particles this difficulty does not arise, and wicking in capillary tubes is the optimal approach with such particles.

6.5.3 Contact angles measured with liquid 1, immersed in liquid 2

Young's equation is not only valid for contact angle measurements of drops of a liquid, L, on a solid, S, in air, but also for the measurement of contact angles of drops of a liquid, L, on a solid, S, immersed in a different liquid which is immiscible in liquid, L, e.g., an oil, O, using:

$$\gamma_{\mathrm{OL}} \cos \vartheta_{OL} = \gamma_{\mathrm{SO}} - \gamma_{\mathrm{SL}} \tag{6.12}$$

(see, e.g., Chaudhury (1984)). It should, however, be noted that with the two-liquid approach, using the advancing contact angle with one liquid implies that one obtains a retreating angle with other liquid. In principle, the two-liquid approach, combined with contact angle measurements in air, can serve to determine $\gamma_{\mathrm{S}}^{\mathrm{LW}}$ values of solids with polar liquids such as water. To that effect one also uses:

$$\gamma_{\mathrm{L}} \cos \vartheta_{L} = \gamma_{\mathrm{S}} - \gamma_{\mathrm{SL}} \tag{6.13}$$

so that subtracting Eq. 6.12 from Eq. 6.13, one obtains:

$$\gamma_{\mathrm{L}} \cos \vartheta_{L} - \gamma_{\mathrm{OL}} \cos \vartheta_{OL} = \gamma_{\mathrm{S}} - \gamma_{\mathrm{SO}} \tag{6.14}$$

For a completely apolar hydrocarbon, O, in view of Eq. 6.2, Eq. 6.14 may be written as in Chaudhury (1984):

$$\gamma_{\mathrm{L}} \cos \vartheta_{L} - \gamma_{\mathrm{OL}} \cos \vartheta_{OL} = -\gamma_{\mathrm{O}} + 2\sqrt{\gamma_{\mathrm{O}}\gamma_{\mathrm{S}}^{\mathrm{LW}}} \tag{6.15}$$

However, when the organic liquid, O, is polar, Eq. 6.14 becomes (see Eq. 6.4):

$$\begin{aligned} \gamma_{\mathrm{L}} \cos \vartheta_{L} - \gamma_{\mathrm{OL}} \cos \vartheta_{OL} = {} & -\gamma_{\mathrm{O}}^{\mathrm{LW}} + 2\left(\sqrt{\gamma_{\mathrm{S}}^{\mathrm{LW}}\gamma_{\mathrm{O}}^{\mathrm{LW}}}\right. \\ & \left. - \sqrt{\gamma_{\mathrm{O}}^{\oplus}\gamma_{\mathrm{O}}^{\ominus}} + \sqrt{\gamma_{\mathrm{S}}^{\oplus}\gamma_{\mathrm{O}}^{\ominus}} + \sqrt{\gamma_{\mathrm{S}}^{\ominus}\gamma_{\mathrm{O}}^{\oplus}}\right) \end{aligned} \tag{6.16}$$

Thus, all three surface tension parameters ($\gamma_{\mathrm{O}}^{\mathrm{LW}}$, $\gamma_{\mathrm{O}}^{\oplus}$, $\gamma_{\mathrm{O}}^{\ominus}$) must be known for the polar organic liquid, still leaving the three unknowns ($\gamma_{\mathrm{S}}^{\mathrm{LW}}$, $\gamma_{\mathrm{S}}^{\oplus}$, $\gamma_{\mathrm{S}}^{\ominus}$) of the polar solid surface (van Oss 1994).

6.6 Other Approaches to the Interpretation of Contact Angle Data

6.6.1 The Zisman approach

Zisman (1964) indicated a correlation between γ_S and the values of γ_L found by extrapolation of $\cos\vartheta$ to $\cos\vartheta = 1$, when ϑ is measured with a number of liquids. The value of γ at $\cos\vartheta = 1$ was termed the *critical surface tension.* However, Zisman was careful never to identify the value for the critical surface tension with the surface tension of the solid, γ_S, as he was well aware of the potential difficulties encompassed in that approach. In the light of Eqs. 6.3 and 7.4 the validity of Zisman's approach holds true only for completely apolar liquids, in which case the determination of γ_S with only one apolar liquid is sufficient, yielding, however, only the value of γ_S^{LW}. As Good (1977) has pointed out, the $\cos\vartheta$ values should be plotted vs. $\sqrt{\gamma_L^{LW}}$, and not γ_L^{LW} as was done in the original Zisman approach (Zisman, 1964). Zisman plots retain a certain utility in the qualitative detection of polar properties of solid surfaces, by showing the unmistakable deviation from linearity among those points that have been obtained with polar liquids. Zisman's pioneering work has been of importance to the general development of characterization methods of solid surfaces in many laboratories (van Oss 1994).

6.6.2 The single polar parameter or "γ^P" approach

There are a number of authors (Owens and Wendt, 1969; Kaelble, 1970; Hamilton, 1974; Andrade *et al.*, 1979; Janczuk *et al.*, 1990) who, probably for intuitive reasons, implicitly postulated the existence of only *one* general property of polar molecules (designated by superscript, P), and who preferred a geometric mean combining rule for the polar interfacial tension, analogous to that used for γ^{LW} (Eq. 5.26) so that:

$$\gamma_{12}^P = \left(\sqrt{\gamma_1^P} - \sqrt{\gamma_2^P}\right)^2 \tag{6.17}$$

the geometric mean combining rule for γ^{LW} is accurate to within 2% (Chaudhury, 1984; van Oss 1994). However, the geometric mean combining rule (Eq. 6.17) is not only erroneous from a theoretical viewpoint (see Chapter 6) but the application of the "γ^P" approach to various practical situations can be spectacularly unpredictive.

The crucial point to note is that when the value of γ_{12}^P is derived from the square of the difference between two entities (following the geometric mean rule; see Eq. 6.18), *it always must have a positive value*, so that ΔG_{121}, which is equal to -$2\gamma_{12}$, then must *always be negative* (cf. Eq. 6.26). Thus, if instead of γ^{AB} (Eq. 6.39), γ^P, as defined in Eq. 6.18, is used, the interaction between two molecules or two particles (1) immersed in a liquid (2) *cannot be otherwise but attractive*, regardless of whether these molecules or particles (or the liquid) are polar or apolar. Especially in the case of macromolecules, such an attraction must either give rise to insolubility in water, or to a very low aqueous solubility (in the limiting case where $\Delta G_{121} \to 0$).

It should be pointed out that if the geometric mean, "γ^P" approach, is followed, Young's equation for polar materials becomes (instead of Eq. 6.4):

$$(1 + \cos\vartheta)\gamma_{\rm L} = 2\left(\sqrt{\gamma_{\rm S}^{\rm LW}\gamma_{\rm L}^{\rm LW}} + \sqrt{\gamma_S^P\gamma_L^P}\right) \tag{6.18}$$

Using the γ^{AB} approach (Chapter 6) and parallel to it, the "γ^P" approach to determine ΔG_{121} in polar media, it is useful to compare the applicability of both of these to the solubility of polymers in water, which is the most polar of liquid media. The solubilities in water (w) of most common water-soluble polymers, including, e.g., poly(ethylene oxide) (PEO), and dextran, using:

$$\Delta G_{\rm SLS} S_c = kT \ln s \tag{6.19}$$

(where S_c is the contactable surface area of the molecule, k is Boltzmann's constant, T is the absolute temperature in degrees Kelvin, and s is the solubility in mole fractions), then becomes exceedingly small (van Oss and Good, 1992; van Oss 1994).

The single polar parameter, or "γ^P" combining rule, applied to polar systems leads to severely erroneous results, and should be avoided in all cases.

6.6.3 The "equation of state"

Neumann's "equation of state" (Neumann *et al.*, 1974) assumes as a matter of principle that the surface tension, and the interfacial tension, must on no account be subdivided into an apolar and a polar component. The computer program pertaining to the "equation of state," in addition, makes it impossible for ΔG to assume a positive value, making it in turn impossible for particle-particle interactions to be repulsive (i.e., for non-electrically charged particles to form stable suspensions) and for neutral polymers (e.g., PEO, polysaccharides) to be soluble in water. Thus, Neumann *et al.*'s "equation of state" is not an appropriate approach for interpreting surface tension data pertaining to polar (e.g., aqueous) systems.

6.7 Contact Angle Determination by Wicking and Thin Layer Wicking

Small, hard particles and powders, with dimensions of the order of 1 μm can of course be fixed in a flat layer, as described in Section 6.5.2, but they then may manifest a "roughness" which invariably results in measured contact angles that are too high (Chaudhury, 1984). Porous materials have the same drawback, as have pressed pellets of particles, with the added difficulty that the contact angle liquid tends to disappear into such porous bodies, by capillarity, before they can be properly measured. However, non-swelling particles, powders or porous solids can still be used in contact angle measurements, by wicking.

Hard particles, of a diameter of 1 μm or more, when spread into flat layers will form a surface that is too rough for accurate contact angle measurement by the sessile (advancing) drop method (Chaudhury, 1984). However, by using a packed column of such particles, capillary rise velocity measurements of a liquid in that column can also yield the contact angle of that liquid with respect to the particles' surface. This "wicking" approach can also be used for contact angle measurement of porous bodies, strands of fibers, etc. In these cases one uses the Washburn equation (Ku *et al.*, 1985; Adamson, 1990):

$$h^2 = \frac{tR\gamma_L \cos\vartheta}{2\eta} \tag{6.20}$$

where h is the height the column of liquid L has reached by capillary rise in time t, R is the average radius of the pores of the porous bed, γ_L is the surface tension of the liquid L, ϑ is the contact angle and η the viscosity of the liquid L.

By first measuring values of h, at a constant time t, for a number of low energy liquids L, which can be taken to spread on the particles' surface (e.g., hexane, octane), the value of R can be obtained for the type of column at hand, packed with a given kind of particle, because in these cases $\cos\vartheta = 1$. Once R is determined, $\cos\vartheta$ can be obtained with a number of well-characterized non-spreading apolar and polar liquids, which then permits the solution of Eq. 6.4. Some doubts could exist whether the Washburn equation 6.20 remains valid even when $\gamma_L < \gamma_S$ (i.e., with spreading liquids). Specifically, it might be questioned whether for spreading liquids the condition $\cos\vartheta = 1$ continues to be applicable. However, it turns out that the liquid pre-wets the surfaces of the particles over which they subsequently spread, in the form of a "precursor film" (de Gennes, 1990; van Oss *et al.*, 1992a).

To use the wicking approach for the determination of $\cos\vartheta$, it is essential to use exceedingly well-packed columns of rather monodisperse particles (Ku *et al.*, 1985). If, upon first contact with an ascending (and lubricating) liquid, tighter packing can occur locally, a gap will be created between particles, which causes a strongly asymmetrical rise of the liquid in the packed column. This makes it very difficult to measure the precise length of travel of the liquid column. However, in those cases where only polydisperse suspensions of irregularly shaped particles are available, an extremely useful alternate approach is the coating of such particles onto, e.g., glass surfaces, followed by the measurement of the capillary rise of various liquids L, as outlined above, in the manner of thin layer chromatography (van Oss *et al.*, 1992a). This method, which was first suggested by Dr. M. K. Chaudhury (private communication), is known as *thin layer wicking.* The thin layer plates are made by deposition of a (2 - 4%) aqueous suspension of the particles or powder onto a horizontal glass slide (e.g., a microscope slide of approximately 1 x 3 inches), and allowing the water (or other suspending liquid) to evaporate. With clay and other mineral particles, the slides can be further dried in an oven, at 105 - 110°C, and should then be kept in a vacuum desiccator until measurement. Thin layer wicking is especially useful in the characterization of non-swelling clay particles (van Oss *et al.*, 1992a; Giese *et al.*, 1996). However, with clay particles that are prone to swelling, wicking should not be used, but swelling particles tend to form very smooth layers upon drying, so that in such cases direct contact angle measurements can be done.

Experimental proof that the $\cos\vartheta$ value obtained by thin layer wicking is the same as the $\cos\vartheta$ value obtainable by *direct* contact angle measurement, was obtained by thin layer wicking of monosized cubical shaped hematite particles, of approximately 0.6 μm. On account of the all around flat sidedness of these particles, the thin layer wicking plates could not only be used for wicking, but they were so smooth that they could also be used for direct contact angle measurements. the results were such that for γ^{LW} 46.1 mJ/m^2 was found by thin layer wicking (TLW) and 45.6 mJ/m^2 by direct advancing contact angle measurement (ϑ_a; similarly $\gamma^{\oplus}$ was 0.1 mJ/m^2 (TLW) and 0.3 mJ/m^2 (ϑ_a), and $\gamma^{\ominus}$ was 50.1 mJ/m^2 mJ/m^2 (TLW) and 50.4 mJ/m^2 (ϑ_a)

(Costanzo *et al.*, 1995).

For the surface tension components and parameters, as well as the viscosities of the liquids used in wicking, see Table 6.1. For wicking, the use of glycerol is not recommended, on account of its very high viscosity. Also, liquids that yield contact angles higher than 90° cannot be used, as they give rise to negative $\cos\vartheta$ values, which prevents capillary rise (cf. Eq. 6.20).

6.7.1 Determination of the average pore radius, R

The success of the thin layer wicking procedure rests on the ability to fabricate a series of glass slides covered with uniform thin layers of finely powdered sample. This is usually possible for materials which form stable or nearly stable suspensions in water (the most common medium for minerals). To ensure that the powder remains in suspension, the liquid should be stirred with a magnetic stirrer during the process. The typical glass microscope slide will support approximately 5 ml of aqueous suspension. The number of slides to be prepared for each sample is determined by the number of liquids used and the number of replicates desired. The replicates (more than one slide with the same liquid) allow an estimate of the uniformity of the sample procedure. The alkanes listed in Table 6.1 can be used to determine the pore radius R (Eq. 6.20). For such spreading liquids, the Washburn equation can be written as:

$$\frac{2\eta h^2}{t} = R\gamma_{\mathrm{L}} \tag{6.21}$$

which is in the form of a straight line $y = mx$ passing through the origin with:

$$y = \frac{2\eta h^2}{t} \tag{6.22a}$$

$$m = R \tag{6.22b}$$

and

$$x = \gamma_{\mathrm{L}} \tag{6.22c}$$

For the wicking coefficient, the unknowns to be measured are h and t. These are determined by measurement of the height of capillary rise, h, at time, t for each spreading liquid. This is best done by marking each slide with small scratches near the periphery of the long dimension of the slide with a fine needle. An accurate ruler is used to place the marks at, e.g., every few

Time	Capillary Rise (t in sec)			
(h in mm)	octane	decane	dodecane	hexadecane
0.2	3	4	13	20
0.4	13	20	41	77
0.6	30	45	83	160
0.8	51	76	135	270
1.0	75	129	210	405
1.2	109	160	286	570
1.4	148	215	373	760
1.6	190	273	492	990
$h^2/t(\times 10^4) =$	1.67	1.92	1.94	2.13
$\gamma_L =$	21.62	23.83	25.35	27.47

Table 6.3: *h and t values for a wicking experiment using spreading alkane liquids with thin layers of a precipitated calcite material. In the bottom line are shown the calculated wicking coefficients based on the data in the table.*

millimeters. These marks do not interfere with the measurement. The sample must be equilibrated with the vapor of each spreading liquid before the wicking experiment can be done. This is easily accomplished by suspending the slide vertically in a weighing bottle (Kimax brand, Kimble No. 15146-3060; 30 mm inside diameter and 60 mm height) of suitable dimensions for the insertion of the microscope slide covered with the powdered sample. For the equilibration, a rubber stopper which fits the weighing bottle is partially cut in half so that it can grasp the end of the glass slide and hold it above the surface of the spreading liquid which reside in the bottom of the bottle. Only a few milliliters of liquid are needed. Once the equilibration is finished (usually 1 hour), the slide is slid down to contact the surface of liquid at which point wicking begins. At this instant, a stop watch is started and the times at which the liquid front reaches each of the scribed marks on the slide are noted. This process yields a series of h and t values. What is needed is the ratio of h^2/t which can be derived from a plot of h^2 versus t values which are fitted to a straight line. The $t = 0$ value is unknown because of the uncertainty of the time when the liquid begins to wick through the powder film; but this is not a problem because the ratio is the desired value, which is the slope of the straight line and this is independent of the $t = 0$ value. As an example, Table 6.3 lists time and height values for four spreading alkanes through thin layers of rhombohedral precipitated calcite, and Figure 6.2 shows a plot of h^2 versus t.

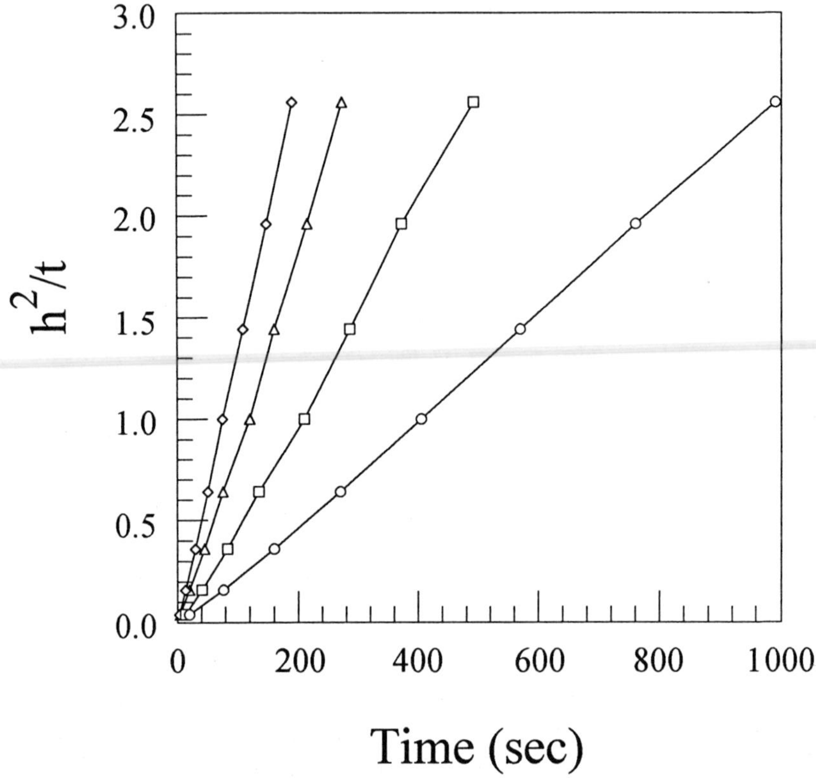

Figure 6.2: *The height, h^2, versus time, t, measurements for the wicking of apolar spreading liquids through a thin layer of calcite particles deposited on glass slides (data from Table 6.3).*

6.7.2 Derivation of contact angles from wicking measurements

Having determined values of h^2/t for the spreading liquids (usually 3-6 liquids), one plots $2\eta h^2 t$, the so-called wicking coefficient, versus γ_L for each and determine the least squares straight line that passes through both the origin and the spreading liquid data points (Figure 6.3). The slope of this line is the average pore radius, R. This is the radius of a capillary tube that would wick a given liquid at the same rate as the powder would. This straight line is in fact the solution of the Washburn equation for $\vartheta = 0$ and $\cos\vartheta = 1$. No experimental points can exist above this line for reasons discussed in an earlier section (6.7.1). Similarly, the field is bounded by the line $h^2/t = 0$; below this line, $\gamma_L < \gamma_S$, i.e., the contact angle is greater than 90° prohibiting wicking (Li *et al.*, 1993). The next step in the analysis is to take the R value determined as above from the spreading liquids, and use this value in the Washburn equation 6.20 to determine $\cos\vartheta$ for each of the non-spreading liquids (both polar and non-polar). This can be done by averaging the h^2/t values for each liquid and inserting these in the Washburn equation, solving for $\cos\vartheta$. The result of the wicking experiments are a list of contact angle (ϑ) values for several non-spreading liquids, ready for analysis via the Young-Dupré equation.

6.7.3 Other uses of wicking

The specific surface area, A, of powders or particles, usually measure by the BET method (see e.g., (van Olphen and Fripiat, 1979; Gregg and Sing, 1982)), can also be determined with thin layer wicking (van Oss *et al.*, 1993). By using low energy, spreading liquids (e.g., heptane, octane, decane, dodecane), the effective interstitial radius, R, (Eq.6.20) can be determined by thin layer wicking. Then by using:

$$R = 2\frac{1 - \Phi}{\Phi \rho A} \tag{6.23}$$

where Φ is the volume fraction of the solid in a packed powder and ρ is the specific density of the individual particles, the specific surface area, A can be determined. With BET and thin layer wicking measurements, specific surface areas were found with talc, several smectites and kaolinite (Table 6.4), that coincided (BET compared with thin layer wicking) within a few percent (van Oss *et al.*, 1993). Also, using (White, 1982):

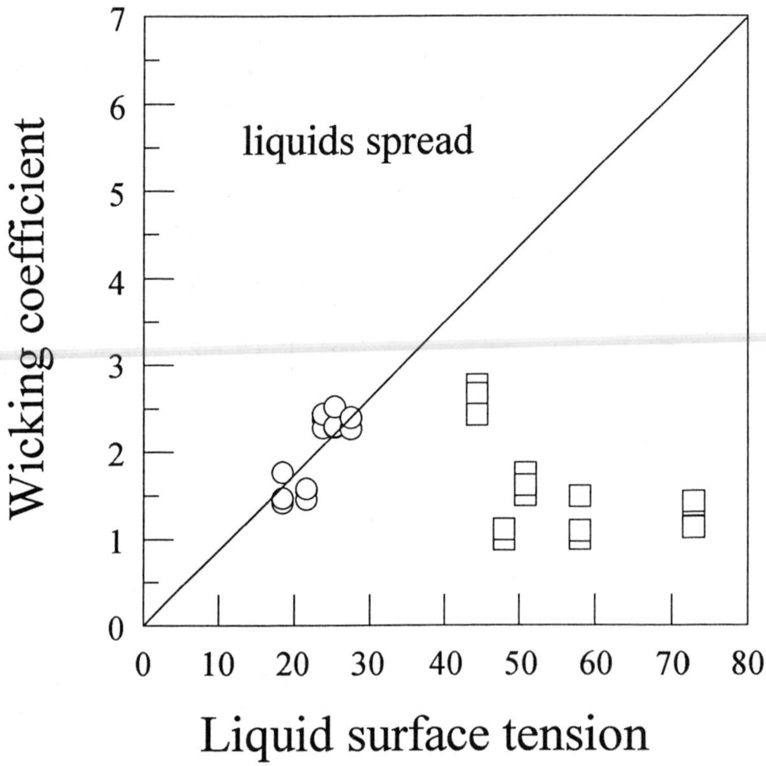

Figure 6.3: *A wicking plot of a talc sample (Li et al., 1993). The wicking coefficient,* $2\eta h^2/t$, *has units of gm cm/sec*^{2} *and the surface tension of the liquid,* γ_L, *has units of mJ/m*^{2} *(or dynes/cm). The slope gives R in cm.*

Mineral	A (m²/gm) BET	Wicking	Φ (cm)	R (10^{-6})	Δ (%)	R_p μm
talc	2.38	2.22	0.299	76.8	+6.7	0.49
SWy-1 (Natural)	23.77	24.59	0.588	2.28	-3.3	0.05
SWy-1 (La)	4.43	5.06	0.619	9.74	-12.5	0.24
KGa-1	9.56	9.46	0.446	10.1	+1.0	0.12
KGa-2	21.76	21.93	0.488	3.68	-0.8	0.05

Table 6.4: *Specific surface areas (A), Φ and R values for several mineral powders. The average sphere radius, R_p, was calculated with Eq. 6.24.*

$$R_p = \frac{3\Phi R}{2(1-\Phi)} \tag{6.24}$$

an approximate value can be obtained for the average particle radius, R_p, assuming uniform spherical particles (van Oss *et al.*, 1993). These values are smaller than the equivalent spherical dimension (e.s.d.) obtained by Stokes law settling in water. For example, montmorillonite, SWy-1, typically has e.s.d. values of a few μm as opposed to 0.05 μm, reflecting the large anisotropy of the smectite particles.

Wicking can also be used for the determination of the average pore size, R, in ceramic materials such as hot pressed powders and sintered gels, using spreading liquids. As an example, Li *et al.* (1994) examined two different silica gels, fabricated from mixtures of colloidal silica and soluble silica (potassium silicate) by wicking with hexane, octane, decane, dodecane and hexadecane. The wicking values of 52.3 ± 0.9 and 32.9 ± 0.9 nm were obtained for 10:90 and 20:80 (i.e., a composition of 20 wt% colloidal silica and 80 wt% soluble silica), respectively. By the usually employed method of mercury intrusion porosimetry (MIP) mean pore sizes were found for these materials of 120 ± 6 and 48 ± 2.4 nm, respectively which were a factor of 2.3, resp., 1.46, higher than the wicking values (Li *et al.*, 1994). Pore sizes down to about $R = 10$ nm can be measured by wicking; however, at even lower pore sizes the rate of wicking becomes so slow that it becomes difficult to measure the capillary rise with precision, with volatile low-energy liquids such as alkanes. In view of the fact that it is almost impossible to safeguard against surface contamination of the very high energy liquid, mercury, the surface tension of mercury (ideally $\gamma_L \approx 480$ mJ/m²) is usually significantly overestimated (see, e.g., Adamson (1990), page 100) so that one tends to find pore sizes via MIP that are too high. Thus, the R-values found by wicking would tend to be closer to the real average R than the porosity determined by MIP (Li *et al.*, 1994).

6.8 Solution of the Young-Dupré Equation from Contact Angle Measurements

6.8.1 Minimal solution

Examination of the Young-Dupré equation (Eq. 6.7) shows that, for the normal application of drops of known liquids on a solid surface of unknown properties, there are three unknowns, γ_S^{LW}, $\gamma_S^{\oplus}$, and $\gamma_S^{\ominus}$. Determination of the values of these unknowns minimally requires three equations, hence contact angles are measured with three different liquids, but two of these liquids *must* be polar. The contact angles for an apolar liquid yields directly the value of γ_S^{LW} by:

$$\gamma_S^{LW} = \gamma_L \left(\frac{1 + \cos\vartheta}{2} \right)^2 \tag{6.25}$$

because $\gamma_L^{\oplus}$ and $\gamma^{\ominus} = 0$. For example, if the contact angle of α-bromonaphthalene on a solid surface is 50.9°, then $\gamma_S^{LW} = 29.5\ \mathrm{mJ/m^2}$. If two or more apolar liquids give non-zero contact angles, each can be used to determine γ_S^{LW} and the results averaged. Having determined γ_S^{LW} from the apolar liquid(s), γ_S^{LW} can be inserted into the Young-Dupré equation leaving only two unknowns, $\gamma_S^{\oplus}$ and $\gamma_S^{\ominus}$. The Young-Dupré equation is non-linear, but it can be linearized easily for the polar unknowns as:

$$a = bx_1 + cx_2 \tag{6.26a}$$

where

$$a = \left(\frac{1 + \cos\vartheta}{2} \right) \gamma_L - \sqrt{\gamma_L^{LW} \gamma_S^{LW}} \tag{6.26b}$$

$$b = \sqrt{\gamma_L^{\oplus}} \tag{6.26c}$$

$$c = \sqrt{\gamma_L^{\ominus}} \tag{6.26d}$$

and

$$x_1 = \sqrt{\gamma_S^{\oplus}} \tag{6.26e}$$

$$x_2 = \sqrt{\gamma_S^{\ominus}} \tag{6.26f}$$

If two polar non-spreading liquids are used, there will be two equations of the type 6.26a and the solution is a simple algebraic exercise.

6.8.2 The overdetermined case

If one uses more than two polar liquids, the number of equations exceeds the number of unknowns, the problem becomes overdetermined and the solution by hand becomes difficult. However, the problem is readily handled in the linearized form by standard least squares analysis where one seeks to minimize the sum of the squared difference between the observed $\cos\vartheta$ values and the values calculated from an initial guess or intermediate refined values.

6.8.3 Estimation of errors in the γ values

First, it should be pointed out that water is the most important of the contact angle liquids. This follows from the very large value of $\gamma_W^{\oplus} = 25.5$ mJ/m^2, by far the largest value of the Lewis acid parameter yet found. Since the $\gamma_L^{\oplus}$ interacts with the $\gamma_S^{\ominus}$, a large $\gamma_L^{\oplus}$ value ensures that the product $\gamma_L^{\oplus}\gamma_S^{\ominus}$ will be large, as will be the contribution to the solution of the simultaneous equations. This yields a well determined value of $\gamma_S^{\ominus}$. In addition, the other polar liquids have $\gamma_L^{\oplus}$ values which are low, and similar to each other, creating difficulties in finding a unique and reliable value of $\gamma_S^{\ominus}$. If a least squares procedure is used for the solution of the simultaneous Young-Dupré equations, omitting water can result in a singular or nearly singular matrix whose inversion becomes difficult and unreliable, yielding γ-values of low reliability (Table 6.5). The origin of this unreliability can be seen when comparing the $\gamma_L^{\oplus}$ values of, in this example, glycerol ($\gamma_L^{\oplus} = 3.92$ mJ/m^2) and formamide ($\gamma_L^{\oplus} = 2.28$ mJ/m^2) both of which are small and similar in magnitude. Similarly, it is possible to determine the values of the three unknowns using three polar liquids. The γ^{LW} values for the apolar liquids are not very different (see Table 6.1) and this leads to an uncertainty in the calculated value of γ_S^{LW} as shown in Table 6.5. Thus, water must be one of the liquids and at least one apolar liquid should be included in order to derive the maximum accuracy from the contact angle data. It is possible to estimate the errors in the calculated γ_S^{LW}, $\gamma_S^{\oplus}$ and $\gamma_S^{\ominus}$ values by a Monte Carlo simulation in which each of the observed contact angle values is randomly modified using the putative standard deviation in the measurement and a Gaussian distribution of the errors. For each set of modified contact angles, surface tension components

Liquid	$\vartheta_{obs.}$	Set A $\vartheta_{calc.}$	Set A $\Delta\vartheta$	Set B $\vartheta_{calc.}$	Set B $\Delta\vartheta$	Set C $\vartheta_{calc.}$	Set C $\Delta\vartheta$
DIM	38.0°	38.06°	0.06°	38.08°	-0.08°	-	-
ABM	24.4	24.24	0.16	24.26	0.14	-	-
GLY	29.3	29.21	0.09	29.31	-0.01	28.9	+0.41
FORM	7.8	8.19	-0.39	7.75	0.05	6.38	+1.42
WATER	3.6	3.63	-0.03	-	-	1.53	2.07
	γ_s^{LW}	40.6	$\pm$ 0.14	40.6	$\pm$ 0.15	41.0	$\pm$ 2.04
	$\gamma_s^{\oplus}$	1.1	$\pm$ 0.03	1.4	$\pm$ 0.20	1.10	$\pm$ 0.34
	$\gamma_s^{\ominus}$	55.5	$\pm$ 0.20	49.8	$\pm$ **5.09**	55.4	$\pm$ 0.33

Table 6.5: *Contact angle measurements on a freshly cleaved muscovite mica surface, each having an assumed standard deviation of 0.5°. The liquids are: DIM = diiodomethane, ABM = α-bromonaphthalene, FORM = formamide, GLY = glycerol, WATER = water. From these angles, surface tension components and parameters (mJ/m²) were calculated using all the contact angle data (Set A), without the water contact angle (Set B) and using only the polar liquid contact angles (Set C). The value of $\gamma^{\ominus}$ becomes very unreliable when the water contact angle is omitted, while the omission of the apolar liquids introduces a moderate uncertainty in γ^{LW}.*

and parameters are calculated; this is repeated some predetermined number of times. The resulting set of γ values can be used to calculate the standard deviation of the distribution of each component and parameter. For the example in Table 6.5 of a freshly cleaved mica crystal, the standard deviations of the observed contact angles are small, on the order of 0.5°. Using the Monte Carlo approach, the standard deviations were calculated for the set with water and without water; these are listed in Table 6.5 and clearly show the dramatic increase in the unreliability of $\gamma_s^{\ominus}$ when the water contact angle is missing.

6.9 Other Methods for Determining Surface Properties

6.9.1 Stability of particle suspensions

In an apolar liquid, when $\Delta G_{131}^{LW} \rightarrow 0$, even partly polar particles should reach their maximum stability, i.e., they should not flocculate. In principle, this occurs when γ_1^{LW} (of particles, 1) is equal to γ_3^{LW} (of apolar liquid, 3). Determination of the value of γ_3^{LW} of the liquid in which a suspension of particles, 1, reaches optimal suspension stability can be done by suspending the particles in question in a series of apolar liquids of varying γ_3-values. The tube in which (with the same number of particles per unit volume of liquid) the particle suspension has reached optimal stability, contains the apolar liquid, 3, for which $\gamma_3 = \gamma_1^{LW}$. The series of apolar liquids can be made up of mixtures of apolar liquids, of varying composition of low and higher γ_3-values. It should, however, be noted that even in mixtures of totally apolar liquids, molecules of the liquid with the lowest surface tension preferentially migrate toward the air interface (van Oss 1994).

Determinations of this type have been done with relatively (but not completely) apolar nylon 6,6 particles, suspended in mixtures of relatively (but by no means completely) apolar liquids, i.e., n-propanol + thiodiethanol (Neumann *et al.*, 1984). A γ_1^{LW} value of 39.6 mJ/m^2 was found for these particles, by that approach, which is fairly close to the best value of 36.4 mJ/m^2, based on many contact angle measurements done with various liquids (van Oss 1994). However, the difference of 3.2 mJ/m^2 between the two approaches is probably significant, and indicative of the influence of the polarity of the particles and of the liquid mixtures used. The two main reasons why this approach should not be used in polar systems are:

1. The distance ℓ between particles at maximum stability is usually of the order of $\ell \approx 5$ to 10 nm, where the rates of decay of ΔG_{131}^{LW} and ΔG_{131}^{AB} obey very different laws (see Chapter 9) so that at $\Delta G_{131}^{TOT} = \max$, at that equilibrium value ℓ, one has multiple unknowns and only one equation.
2. Mixtures of polar solvents have especially unreliable surface tensions for use in Young's equation because one of the two polar components invariably orients toward the liquid-air interface, so that the measurable liquid surface tension is no longer proportional to the energy of cohesion of the liquid (Good and van Oss, 1983; Docoslis *et al.*, 2000a). More recently Fuerstenau *et al.*(1991) and Diao and Fuerstenau (1991)

attempted to ascertain the γ-value of the solid surfaces of particles by flotation experiments in liquids of different surface tensions. This approach is analogous to the particle stability experiments described above and has the drawbacks, for the same reasons. In summary, the particle suspension method can only be recommended when really apolar liquids are used, i.e., solely for the estimation of the γ_1^{LW}-value of particles, 1, and even then the use of mixtures of apolar liquids introduces a degree of uncertainty on account of the inhomogeneity of the liquid mixture at the liquid-air and also at the liquid-solid interface.

6.9.2 Advancing freezing fronts

This method utilizes conditions where $\Delta G_{132}^{TOT} \rightarrow 0$, in order to ascertain the γ_1 of solid particles (1), immersed in a liquid (3), which is gradually being solidified into a solid (2) by cooling. Clearly, when such an advancing solidification front (2) pushes particles (1) in front of it, $\Delta G_{132}^{TOT} > 0$, and on the other hand when particles (1) are engulfed by the front as it advances, $\Delta G_{132}^{TOT} < 0$.

In apolar systems, where $\Delta G_{132}^{TOT} = \Delta G_{132}^{LW}$, we can use the Dupré equation (Eq 6.2):

$$\Delta G_{132}^{LW} = \gamma_{12}^{LW} - \gamma_{13}^{LW} - \gamma_{23}^{LW} \tag{6.27}$$

in which ΔG_{132}^{LW} becomes equal to zero when either $\gamma_1^{LW} = \gamma_3^{LW}$ or $\gamma_2^{LW} = \gamma_3^{LW}$. In other words, in apolar systems, when particles (1) immersed in a liquid (3) are neither pushed nor engulfed by the advancing freezing front of the solidifying melt (2), $\gamma_1^{LW} = \gamma_3^{LW}$, which, in principle, is another method for determining the γ_1^{LW} of particles (this presupposes that $\gamma_1^{LW} \neq \gamma_2^{LW}$, which is usually the case). The method was first tested by Omenyi and Neumann (1976) and by Neumann *et al.* (1979) and described in detail by Omenyi (1978); see van Oss (1994). In aqueous media the method can be to determine whether particles are quite hydrophilic, or frankly hydrophobic, by slowly freezing the particle suspension in a test tube, starting by strongly cooling the tube at the bottom, and allowing the column of ice to advance slowly to the top of the tube. For instance, the montmorillonite clay, SWy-1 ($\gamma^{LW} = 42$, $\gamma^{\oplus} = 0.008$, $\gamma^{\ominus} = 30.2$ mJ/m^2), which is hydrophilic, can be made hydrophobic by treatment with hexadecyl trimethyl ammonium (HDTMA) cations (see Section 7.4.1) ($\gamma^{LW} = 41$, $\gamma^{\oplus} = 0.25$, $\gamma^{\ominus} = 12.4$ mJ/m^2). When suspended in water, the hydrophilic SWy-1, upon freezing in a tube, from the bottom, SWy-1 is partly engulfed and partly pushed to the top of the ice in the tube, because of the repulsion between the ice front and the SWy-1 particles, suspended in cold water, i.e., $\Delta G_{132} > 0$. The hydrophobic HDTMA-SWy-1 on the other hand is completely engulfed and remains in the lower 24% of the ice column after final freezing, i.e., for HDTMA-SWy-1; $\Delta G_{132} < 0$ (van Oss *et al.*, 1992c). Admixture of, e.g., 50% glycerol to the water can reverse the rejection of the hydrophilic particles by the advancing freezing front to engulfment. This was demonstrated with erythrocytes in water and in 50% glycerol (van Oss *et al.*, 1992b). (For the freezing of blood cells, repulsion by an advancing ice front leads to osmotic stress and thus to destruction; glycerol is therefore added to the freezing medium to obviate that problem).

Even though one could use various glycerol concentrations to water, to arrive at a point where slightly less glycerol still leads to particle rejection and slightly more glycerol gives rise to inclusion, and supposing one could with precision determine the γ_3^{LW}, the $\gamma_3^{\oplus}$ and the $\gamma_3^{\ominus}$-values of the liquid mixture (which is fraught with uncertainties), one could still only learn the one value at which ΔG_{132} of the system equals zero. This only yields one equation, with the three values (γ_1^{LW}, $\gamma_1^{\oplus}$, $\gamma_1^{\ominus}$) of the particles as unknowns. (When the γ-values for cold water, at 0°C are: $\gamma_3^{LW} = 22.8$, $\gamma_3^{\oplus} = 26.5$, $\gamma_3^{\ominus} = 26.5$ mJ/m^2, the γ-values for ice are, at 0°C: $\gamma_2^{LW} = 29.6$, $\gamma_2^{\oplus} = 13.0$, $\gamma_2^{\ominus} = 28.0$ mJ/m^2, cf. van Oss *et al.* (1992b, c).

Thus, the advancing (aqueous) freezing front method, using water, can only serve to get a rough estimate as to the hydrophilicity (rejection by the ice) or hydrophobicity (engulfment by the ice) of particles, which is probably more easily ascertained by observing whether there is suspension stability, or flocculation, see the preceding Section 7.8.1.

6.9.3 Force balance

Force balances which measure, with great precision, the interaction forces between two solid objects, as a function of distance between the two objects, were developed, independently, in Russia and in the Netherlands, especially during the 1950's (van Oss 1994). Their main initial aim was the experimental verification of the existence of retardation (Casimir and Polder, 1948) of attractive van der Waals forces at distances greater than approximately 10 nm. Experimental proof of the existence of retardation was finally obtained with an improved force balance (Tabor and Winterton, 1969). The usefulness of the force balance was subsequently vastly improved by Israelachvili and Adams (1978), who extended its application to measurements in liquids (e.g., water), whilst theretofore measurements usually were only done in vacuo. The modern force balance (Israelachvili, 1991) uses two crossed quartz half cylinders, coated with a layer of molecularly smooth sheets of muscovite mica, the whole immersed in water. Distances between the crossed cylinders can be measured (optically) within about 0.1 nm.

With this force balance, repulsive forces were measured between uncoated mica surfaces (Pashley, 1981; Pashley and Israelachvili, 1984a,b). These repulsive forces exceeded the repulsion that would be generated by electrostatic forces (caused by the strong negative charge of mica) alone. By coating the mica surfaces electrostatically by means of a quaternary ammonium base comprising hexadecyl groups, the surfaces could be made (partially) hydrophobic (Pashley *et al.*, 1985), but complete, molecularly smooth hydrophobization was not achieved until 1995 by Wood and Sharma, who achieved complete and smooth hydrophobization of the mica half cylinders (Wood and Sharma, 1995). They also showed that with robust and integral coverage the characteristic length of water is not more than 1.0 nm for hydrophobic attraction, i.e., the same value that was earlier established for hydrophilic repulsion (cf. e.g., Christenson (1992)); see Chapter 9.

With the force balance, ΔG_{1w1} and ΔG_{1w2} can be determined fairly accurately at distances $\ell > \ell_0$ (ℓ_0 is the minimum equilibrium distance, at "contact"; $\ell_0 \approx 1.57$ Å, cf. (van Oss 1994), Chapter XI). However, at $\ell = \ell_0$, i.e., at contact, force balance measurements tend to be less precise than, e.g., contact angle measurements which are, *ipso facto*, always most relevant at contact.

6.9.4 Electrophoresis in monopolar organic solvents

The electrophoretic mobility of polar particles in (monopolar) organic solvents is a measure of the particles' polar parameter of the opposite sign of that of the solvent (Labib and Williams, 1984, 1986; Fowkes, 1987). However, the correlation between the polar parameters found for various compounds by this approach is as yet more qualitative than quantitative (van Oss 1994); see also Chapter 8.

6.10 Surface Tension Measurement of Liquids

There exists an extremely useful compendium of published surface tensions of many (mainly organic) liquids, measured by various methods, at different temperatures (Jasper, 1972). These values are usually extremely reliable; only one error has been noted which is of some importance to the present work, i.e., for diiodomethane, at 20°C, γ_L = 50.8 mJ/m^2, and not 66.98 mJ/m^2 (Chaudhury, 1984).

6.10.1 The Wilhelmy plate method

The total surface tension of liquids (γ_L) is most easily determined by the Wilhelmy plate method (Adamson, 1990). Briefly, a thin rectangular plate (e.g., a chromic acid cleaned glass microscope cover slip, or an acid cleaned platinum plate) is suspended vertically from the arm of a microbalance, until it just touches the hanging plate. Upon contact with the liquid, a small additional force, or additional weight (ΔW), is exerted on the plate, so that (Adamson, 1990):

$$\gamma_L \cos \vartheta = \frac{\Delta W}{P} \tag{6.28}$$

where P is the periphery of the plate.

6.10.2 Pendant drop shape

Another rather reliable approach to measuring γ_L is the hanging drop method (Adamson, 1990). After measuring the dimensions d_e and d_s (the maximum diameter d_e, and the diameter d_s measured a distance d_e above the bottom of the drop) from enlarged photographs of the drop, γ_L can be obtained as follows (Adamson, 1990):

$$\gamma_L = \frac{\Delta \rho g d_e^2}{H} \tag{6.29}$$

where $\Delta\rho$ is the difference in density between liquid and the gas phase (or air) and g the acceleration caused by gravity. H is a shape-dependent parameter; Adamson (1990) gives tables, correlating $S = d_s/d_e$ versus $1/H$.

6.10.3 Interfacial tension between immiscible liquids

Interfacial tension (γ_{12}) between immiscible liquids can, in principle, be measured by the same methods with which liquid/vapor surface tensions are measured (Adamson, 1990). However, especially for the measurement of fairly small γ_{12} values, a few of these methods are better adapted for this particular purpose, e.g., the pendant drop-shape method (see above), the simpler, but less accurate drop-weight method (Adamson, 1990), and a few other methods using deformed interfaces, such as the rotating drop method (Vonnegut,

1942; Princen *et al.*, 1967), and the method described by Lucassen (1979), using drops suspended in another liquid with a density gradient. Vonnegut's basic equation is:

$$\gamma_{12} = \frac{1}{4}\omega^2 \Delta\rho r_0^3 \tag{6.30}$$

where ω is the speed of revolution around the long axis, $\Delta\rho$ the density difference and r_0 the maximum radius of the drop at the axis of rotation.

For very small values of γ_{12}, Vonnegut's (1942) rotating drop (or cylinder) method (which has been further elaborated upon by Princen *et al.* (1967)) is the most suitable. As a first approximation, the presence of a straight, flat meniscus at the interface between two liquids inside a vertical tube is an indication of a very low interfacial tension between two liquids. All the above methods are also applicable to measuring γ_L, using hanging drops, in air.

It should be noted however that all these drop-shape or drop-weight methods for measuring *interfacial tensions between two immiscible liquids*, are only valid when applied. e.g., to one polar liquid (such as water) and a *completely apolar* liquid, e.g., octane. This approach to measuring interfacial tensions between water and a *partly polar* liquid, e.g., octanol, yields grossly underestimated γ_{12} values as a consequence of the orientation of the polar molecules moieties of the partly polar liquid, e.g., the OH groups, to the water interface. A much more accurate approach to measuring γ_{12} in such cases then is through the solubility of 1 in 2 (van Oss *et al.*, 2001a).

6.10.4 Apolar and polar surface tension component liquids

γ_L^{LW} is best determined by contact angle measurement of liquid L on a purely apolar solid surface, of $\gamma_S = \gamma_S^{LW}$:

$$(1 + \cos\vartheta)\gamma_L = 2\sqrt{\gamma_S^{LW}\gamma_L^{LW}} \tag{6.31}$$

Once γ_L^{LW} is known, as well as γ_L, γ_L^{AB} can be derived from:

$$\gamma_L = \gamma_L^{LW} + \gamma_L^{AB} \tag{6.32}$$

Apolar solids which have been used for this purpose are polyethylene and polypropylene (Fowkes, 1987) and polytetrafluoroethylene (van Oss *et al.*, 1989b).

	Solid			
	γ_S^{LW}	$\gamma_S^{\ominus}$	$\gamma_G^{\oplus}$	$\gamma_F^{\oplus}$
PMMA	43.2	22.4	4.18	1.72
PMMA	41.4	12.2	4.94	3.24
Clay	39.9	21.5	2.56	1.28
PEO	45.9	58.5	4.29	2.88
	33.0	11.1	2.39	
CPPL	32.7	6.2	4.56	
	33.0	9.1	2.94	
Agarose	41	26.9	5.32	
Zein	41.1	18.7	3.54	
Cellulose acetate	38	32.3	3.78	
HSA	41	20.3	4.62	
	γ_S^{LW}	$\gamma_S^{\ominus}$	$\gamma_S^{\oplus}$	
Nylon	36.4	21.6	0.02	2.8

Table 6.6: $\gamma_L^{\oplus}$ *values for glycerol (G) and formamide (F) derived from contact angle measurements on monopolar basic solids (van Oss et al., 1990b; van Oss 1994).*

An entirely different approach to measuring γ_L^{LW} and γ_L^{AB} (and even $\gamma_L^{\oplus}$ and $\gamma_L^{\ominus}$, see below) is to encase the liquid inside a gel, which allows one to treat the liquid as a *solid*, upon a surface of which drops of another liquid can be deposited, for contact measurement. The interior of a gel, especially when made at a low concentration of the gel-forming matrix, retains to a large extent the principal physico-chemical properties of the continuous liquid phase (Michaels and Dean, 1962). By using this approach, the γ_L^{LW} value of water could be determined, using agarose gels. When an aqueous agarose gel, consisting of X% agarose, is cut open, is cut open, the cut surface may be considered as a flat surface (100 - X)% "solid" water, at room temperature. By measuring contact angles with hexadecane ($\gamma_L = \gamma_L^{LW} = 27.5$ mJ/m^2 at 20°C) on these hydrous surfaces, formed with 2.0, 1.75, 1.50, 1.25 and 1.0% (w/v) agarose gellified with pure water, by extrapolation to 0% agarose, a value of 21.4 ± 1.2 mJ/m^2 was found for the γ_L^{LW} of water (van Oss *et al.*,1987c). The value for γ_L^{LW} found for water in this manner agrees well with the value of 21.8 ± 0.7 mJ/m^2 calculated by Fowkes (Fowkes, 1963) from the interfacial tensions between water and various hydrocarbons given by Girifalco and Good (1957); see also van Oss (1994).

6.10.5 Determination of the polar surface tension parameters of liquids

Next to water, glycerol and formamide are the highest energy hydrogen-bonded polar (non-toxic) liquids readily available for contact angle measurements on polar surfaces. Glycerol (G) has been used, together with water, to obtain the $\gamma_S^\oplus$ and $\gamma_S^\ominus$ values for a variety of polar surfaces (van Oss *et al.*, 1986, 1987a,b). However, a drawback of glycerol is its high viscosity (i.e., 1490 times that of water, at 20°C), which complicates standardization of the timing of contact angle measurement (an important consideration, particularly when there is any degree of solubility between solid and liquid, or where there is some sort of reaction between the solid and liquid, such as intercalation in smectites and vermiculites). Another liquid with a surface tension that is almost as high as that of glycerol is formamide (F; γ_F^{LW} = 58 mJ/m^2, $\gamma_F^\oplus$ = 39 mJ/m^2 and $\gamma_F^\ominus$ = 19 mJ/m^2) (Chaudhury, 1984). Formamide has a viscosity which is much more comparable to that of water (i.e., ≈ 4 centipoises at 20°C) and, like glycerol, formamide also has much more pronounced (Lewis) base characteristics than water, see Table 6.1. Thus, while formamide still has a sizable $\gamma_F^\oplus$ parameter, its $\gamma_F^\oplus/\gamma_F^\ominus$ ratio is sufficiently different from that of water to yield useful contact angle data, which in conjunction with contact angle data obtained with water, will yield reliable values for $\gamma_S^\oplus$ and $\gamma_S^\ominus$ for a variety of polar solids. Also, whilst $\gamma^\oplus$ and $\gamma^\ominus$ can be entirely defined with two polar liquids, the availability of a third polar liquid can yield a useful set of control values. In addition, a fourth polar liquid, ethylene glycol, has been characterized using its γ^{LW} and γ^{AB} values (Chaudhury, 1984), and its solubility data (Table 6.1) (van Oss, 1994).

The $\gamma_W^\oplus/\gamma_G^\oplus$ and $\gamma_W^\ominus/\gamma_G^\ominus$ ratios have been determined by contact angle measurements on surfaces of a number of relatively strong Lewis bases, so as to be able to observe a significant degree of interaction between these base parameters, and the relatively minor acid parameter of glycerol and formamide. The monopolar (Lewis) basic solid surfaces used were poly (methylmethacrylate) (PMMA), poly (ethylene oxide) (PEO), smectite clay films, corona-treated poly (propylene) (CPPL), dried agarose gel, dried zein (a water-insoluble corn protein), cellulose acetate film, dried films of human serum albumin (HSA). The $\gamma_L^\oplus$ values found for glycerol and formamide are given in Table 6.6. The average values for $\gamma_G^\oplus$ and $\gamma_F^\oplus$ derived from the results shown in Table 6.6, are: $\gamma_G^\oplus$ = 3.92 ± 0.7 (SE) mJ/m^2 and $\gamma_F^\oplus$ = 2.28 ± 0.6 (SE) mJ/m^2 From the known γ^{AB}-values of glycerol and formamide, it follows that $\gamma_G^\ominus$ = 57.4 mJ/m^2 and $\gamma_F^\ominus$ = 39.6 mJ/m^2 (cf. Table 6.1).

7

Electrokinetic Methods

While the electrokinetic surface, or ζ-potentials, originate from the surface or interfacial properties of solid materials, they are actually situated about 0.3 to 0.5 nm outside a material's surface and have to be extrapolated inward to the (ψ_0) potential at the actual surface, using Eq. 5.54. The electrostatic free energy of interaction, ΔG^{EL}, between two surfaces, 1, reaches a value of about +1.0 mJ/m^2 at $\psi_0 \approx 75$ mV, in an aqueous medium with a 100 mM salt content of a mono-mono-salt; see Table 5.1. Now various clay and other mineral particles can have ψ_0-potentials that are between 50 and 90 mV, in which case ΔG^{EL}, while not dominant, is no longer negligible. For instance for a contact between two platey clay particle surfaces over about 100 nm^2 (= 10^{-10} cm^2) an attraction of 1 mJ/m^2 still corresponds to $\Delta G^{EL} \approx 2{,}500$ kT. Thus, it is always wise to measure ζ-potentials, from which the actual surface, or ψ_0-potential can be derived.

7.1 Electrophoresis

Electrophoresis is by far the most convenient of the electrokinetic methods for measuring the ζ-potential of molecules, macromolecules and particles. For diagrams of electrophoresis and other electrokinetic modes, see Figure 7.1. In the context of this work, particle electrophoresis is the most important electrokinetic mode for measuring ζ-potentials. Mineral particles for which the ζ-potentials are of interest, are usually of the order of one or a few μm in diameter, whereas the thickness of the diffuse double layers surrounding them in most cases only vary between 0.1 and 10 nm, the equation connecting the electrophoretic mobility (μ) of such particles, and their ζ-potential, in all

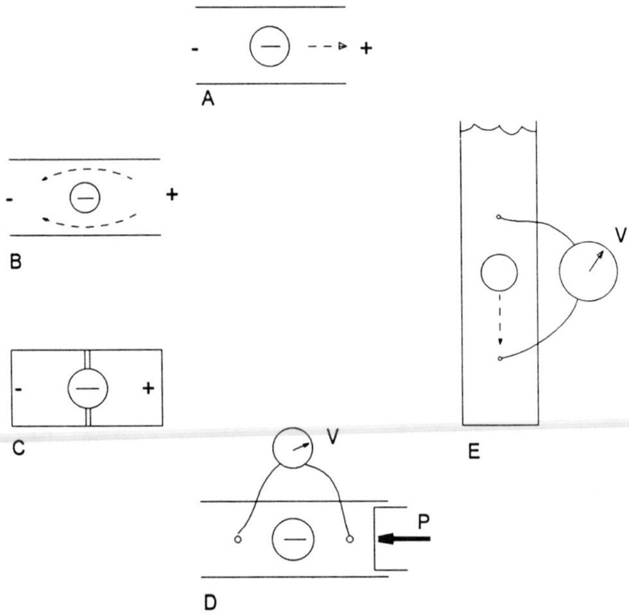

Figure 7.1: *Schematic representation of five electrokinetic phenomena: A) electrophoresis, B) electroosmosis, C) electroosmotic counter-pressure, D) streaming potential, and E) migration potential; for details see text. In a clockwise as well as counterclockwise direction each one of these five electrokinetic phenomena depicted above stands in a simple relationship to its closest neighbors: "-" and "+" indicate the polarity of an applied electric field; "P" indicates pressure; "-" stands for a particle with a negatively charged surface; "——" indicates that such a particle is immobilized by attachment inside the chamber; and "V" designates a voltmeter indicating an electrical potential difference resulting from the last two electrokinetic phenomena (D and E). Not drawn here is a model of streaming current, which is closely related to streaming potential and has analogies with electroosmosis (van Oss 1979, 1994).*

cases is Eq. 5.62:

$$\mu = \frac{\zeta\varepsilon}{4\pi\eta} \tag{5.62}$$

The method of choice for measuring the electrophoretic mobility, μ, of minerals particles is microelectrophoresis, treated below.

7.1.1 Particle microelectrophoresis

The simplest kind of microelectrophoresis is usually effected inside a capillary or chamber which is closed at both ends. In such systems the electroosmotic backflow along the walls of the capillary or chamber, in conjunction with the forward electrophoretic movement of the particles, results in a parabolic particle migration profile, so that the true particle electrophoretic mobility is to be found only at the precise level of the "stationary layer" (Seaman and Brooks, 1979), which necessitates very precise focusing of the microscope. This causes severe problems with microelectrophoresis of larger particles, or high density particles, such as clay particles, which fairly quickly sediment away from the stationary layer and thus tend to get out of focus, or disappear altogether from view. This effect may to some extent be countered by coating the capillary wall with a material of low ζ-potential (Seaman and Brooks, 1979), or preferably by covering the capillary wall with a thin gel of low ζ-potential, inside of which the electroosmotic backflow is trapped, thus leaving the lumen free of any electroosmotic flow gradient, permitting one to focus at all levels inside the lumen left by the gel (van Oss *et al.*, 1974; van Oss and Fike, 1979).

Automated microelectrophoresis has been developed using laser Doppler velocimetry (van Oss and Fike, 1979) and related methods. The costs of these various approaches differ widely; automated devices' prices are of the order of $\$10^5$; stationary layer instruments cost about $\$2 \times 10^4$ and gel-coated capillary microelectrophoresis devices $\$2$ to 3×10^3. In our own laboratory we have found it expedient always to have a gel-coated capillary microelectrophoresis available as a stand-by. The operation of this simple device is somewhat more laborious, but it never breaks down, which cannot always be said of the more automated apparatus.

7.1.2 Electrophoresis in non-aqueous media

Electrophoresis of mineral particles in non-aqueous media is rarely utilized. It could however, in principle, yield some information on the $\gamma^{\oplus}$ or $\gamma^{\ominus}$-values of mineral particles, as discussed in Section 6.9.4 (see also van Oss (Chapter X, pp 152-153) 1994).

7.2 Electroosmosis

It has been long established that there essentially is total equivalence between the electrokinetic phenomena (see Figure 7.1) and that ζ-potentials can theoretically be obtained through electroosmosis or streaming potential as well as through electrophoretic determinations.

Synge and Tiselius (1950) and Mould and Synge (1954) first proposed to make use of electroosmotic flow in gel membranes or other porous media, to transport solvents through such porous media and thus to obtain separation according to the site of solutes transported with that solvent, the smaller solvent molecules being able to go through smaller pores than the larger ones (see above). In the 1970s, Pretorius *et al.* (1974) used electroosmosis as a means of transporting solvents, in thin-layer and in high-speed chromatography. The advantages of electroosmosis over other means of transporting solvents are two-fold: (1) with a reasonably high ionic strength (obtained by addition of electrolytes), and thus with a compressed electric double layer (see Chapter 5), virtually ideal plug-flow can be achieved throughout the column, with a concomitant reduction in theoretical plate height (Pretorius *et al.*, 1974); (2) In open-ended tubes electroosmosis generates no noticeable hydrostatic pressure, and the rate of solvent transport is relatively independent of the size of the pores or channels. van Oss and colleagues have used electroosmosis and electrophoresis combined and in the same direction through stacks of dense membranes to achieve enrichment of ^{7}Li (van Oss *et al.*, 1959). Doren *et al.* (1989) used electroosmosis to determine the ζ-potential of flat plates, in a modified microelectrophoresis device.

Thus, electroosmosis can be used as a means of solvent transport and as a method for determining the surface potentials of flat solid plates.

7.3 Streaming Potential and Sedimentation Potential

ζ-potentials of minerals and other granular materials can be determined by streaming potential measurements (Fuerstenau, 1956; Somasundaran and Kulkarni, 1973); such measurements are of considerable importance in the mining industry because of a strong correlation between the flotation properties of minerals and their ζ-potentials (Somasundaran, 1968, 1972). Other devices for the study of streaming potentials have been developed by Rutgers and de Smet (Rutgers and de Smet, 1947) and by Boumans (1957) who by means of turbulent flow in metal capillaries used streaming potentials to build up extremely high voltages (70,000 V) in a fluid-flow type of van de Graaf generator, developing, however, only 10^{-7} amps. It is well known however that in pipes through which low-conductivity liquids, such as petroleum, gasoline, etc., are pumped, potential of thousands of volts can be built up, giving rise to the formation of sparks, which cause explosions (Boumans, 1957). Tuman (1963) measured streaming potentials through sandstone (up to several V) at very high pressures (up to 2,000 p.s.i.). Improvements on existing pressure ζ-potential measuring methods have been discussed by Korpi and de Bruyn (1972). van Oss (1963) and van Oss and Beyrard (1963) described the role of streaming potentials elicited by the potential difference used in ultrafiltration through charged membranes, on the ion retention by these membranes.

The school of Somasundaran has done much work on the development of streaming potential measurement methodology (Somasundaran and Kulkarni, 1973), and on changes occurring in various minerals surfaces, as a function of time (Somasundaran, 1968), pre-treatment (Somasundaran, 1972) or conditions of pH or ionic contents of the aqueous phase, using beryl, calcite and apatite.

Sedimentation potentials elicited by the settling of charged particles under the influence of a gravitational field (Dorn effect) have been little studied by direct methods. The influence of sedimentation potentials on the sedimentation coefficients of polyelectrolytes has been studied by Mijnlieff (1958). Flotation potentials caused by gas bubbles rising through a liquid are somewhat easier to measure; determinations were done at the beginning of this century by McTaggart (1914a, b, 1922) and were continued by Alty (1924, 1926). More recently Usui and Sasaki (1978) reported on bubbles through aqueous solutions of cationic, anionic and nonionic detergents. The potentials of rising air bubbles were taken into account in flotation experiments done with nitrocellulose by LaFrance (1994).

7.4 Link Between the Electrokinetic Potential and Electron Donicity

7.4.1 The Schulze-Hardy rule

H. Schulze in the 1880's and W. R. Hardy in 1900 observed that for the onset of flocculation of, e.g., a dilute negatively charged particle suspension in water one needs roughly 10^2 mM monovalent cations (counterions), 1 mM divalent cations and 10^{-2} mM trivalent cations (Overbeek, 1952). This phenomenon has been traditionally ascribed to the effect of these counterions to depress the ζ-potential of the particles in suspension, which then causes the van der Waals attraction between the particles to prevail, which gives rise to flocculation. A closer quantitative analysis of the changes in ΔG^{EL}_{1w1} shows, however, that in many cases of flocculation due to a decrease in ζ-potential, the decreased but still positive value of ΔG^{EL}_{1w1} is still greater than the negative value of ΔG^{LW}_{1w1}, even though pronounced flocculation definitely takes place (Wu *et al.*, 1994a). A third force acting in water therefore must play a decisive role. That third force is a net Lewis acid-base (AB) attraction (see Chapters 6 and 9), occurring as a result of the decrease in ζ-potential, which can be caused by the admixture of plurivalent counterions, or by a change in pH.

The first report of this effect was by Ohki (1982), who noted, by contact angle measurements with water on negatively charged phospholipid surfaces that an increased Ca^{2+} content of the water drops resulted in an increase in contact angle, i.e., in an increase in hydrophobicity. Holmes-Farley *et al.* (1985) measured contact angles with drops of water at different pH's, on amphoteric polymer surfaces and observed that the closer the pH was to the isoelectric point of the amphotere, the higher the contact angle. In 1988 van Oss and Good (1988) observed that human serum albumin becomes hydrophobic at its isoelectric pH of 4.8; see also van Oss (1994).

The hydrophobizing effect of a decrease in ζ-potential on suspensions of mineral particles was more thoroughly studied by Wu *et al.* (1994a,b) and by Wu (1994). A typical example is given in Table 7.1, where a very stable aqueous suspension of particles of ground glass is studied as is, and after the admixture of 0.47 mM $LaCl_3$ or of 2.4 mM $CaCl_2$. In both cases the admixture of these counterions caused a significant decrease in ζ-potential, from -52.7 mV to -16.4 mV (with La^{3+}) and to -14.0 mV (with Ca^{2+}). The values of ΔG^{EL}_{1w1} thus decreased from +3,000 kT to +770 and +540 kT respectively. This of course is a strong decrease in ΔG^{EL}_{1w1}, but it does not suffice to allow

	A	B	C
$LaCl_3$ (mM)	0	0.47	0
$CaCl_2$ (mM)	0	0	2.4
ionic strength	15	17	21
$\Gamma/2\times10^{-3}$	15	17	21
pH	7.5	7.6	7.4
ζ(mV)	-52.7	-16.4	-14.0
Stability	++	–	–
γ_s^{LW} (mJ/m^2)	31.5	30.3	29.6
$\gamma_s^{\oplus}$ (mJ/m^2)	0.4	0.2	0.3
$\gamma_s^{\ominus}$ (mJ/m^2)	37.1	20.9	22.2
ΔG_{1w1}^{LW} (kT)	-220	-170	-145
ΔG_{1w1}^{AB} (kT)	+14,000	-6,700	-4,700
ΔG_{1w1}^{EL} (kT)	+3,000	+770	+540
ΔG_{1w1}^{TOT} (kT)	+17,000	-6,100	-4,300

Table 7.1: *Surface properties of ground glass particles, 0.5% (w/v) in 20 ml, in various suspensions in PBS/10 ($\Gamma/2$ = 0.015) untreated* (A)*and treated with La^{3+}* (B) *or Ca^{2+}* (C)*. A "+" indicates stability while "-" indicates instability. From Wu (1994); see also Wu et al. (1994a).*

the van der Waals attraction, ΔG_{1w1}^{LW}, of respectively -170 and -145 kT to predominate and to cause flocculation. The principal cause of the change from stability to flocculation is ΔG_{1w1}^{AB}, which veered from +14,000 kT to -6,700 kT (La^{3+}) and -4,700 kT (Ca^{2+}). In other words, the decrease in ζ-potential of these negatively charged particles under the influence of La^{3+} and Ca^{2+} ions, caused a dramatic change from a hydrophilic (AB) repulsion of +14,000 kT to a hydrophobic (AB) attraction of -6,700 and -4,700 kT, respectively.

This hydrophobization, linked to a decrease in the negative value of the ζ-potential by the admixture of small amounts of plurivalent counter ions, has been shown to cause flocculation of a variety of initially stable particle suspensions, e.g., calcite powder, montmorillonite (SWy-1) particles (Wu, 1994; Wu *et al.*, 1994a,b). The flocculated particles could be reconverted to a stable suspension by complexation of the di- and trivalent counterions with, e.g., EDTA or hexameta phosphate (Wu, 1994). Upon standing for about 16 hours, the flocculated suspensions repeptized spontaneously, as a result of intercalation of the counterions into the clay particle structure (Wu, 1994).

8

Interactions Between Colloids

8.1 Introduction

When two condensed materials are in intimate contact across an interface, the surface atoms of each are able to interact across the interface. Any understanding of the interfacial free energy or the interfacial tension ultimately rests on an accurate understanding of the types of interactions which exist across the interface. Part of this interaction is electrodynamic (i.e., pertaining to van der Waals interactions) and has been recognized for a century or more (see Chapter 5). The recognition of other types of interaction has occurred more recently, the major one involving the non-covalent sharing of electrons across the interface. This non-covalent bonding is part of the Lewis acid-base (AB) description which is widespread in modern chemistry.

8.2 Lifshitz-van der Waals Interactions

The theory of van der Waals forces is treated extensively in Chapter 5. It should be noted here that attractive Lifshitz-van der Waals (LW) interactions between adjoining flat parallel plates *in vacuo* are quite strong; the (negative) value for $\Delta G_{\mathrm{ss}}^{\mathrm{LW}}$ between two such solids *in vacuo* can reach 70 to 100 mJ/m^2 (e.g., clay platelets). However, in aqueous media such attractive LW energies decrease to 3.1 to 11.5 mJ/m^2. Nonetheless, depending on the system under study, even the latter values may not be negligible.

8.3 Electrostatic Interactions

The theory of electrostatic interactions is treated in Chapter 5. Similarly to AB-forces and in contrast with LW-forces, EL-potentials have two opposite signs; both of these may be, *microscopically*, present on a given solid surface, but only the dominant sign of charge of these can be *macroscopically* measured.

It should also be noted that there exists a linkage between EL and AB interactions. For instance, a decrease in ζ-potential (cf. Chapter 5), which causes a decrease in EL-repulsive energies, but at the same time gives rise to a much stronger decrease in AB-repulsive energies, usually even causing an AB-repulsion to become an AB-(hydrophobic) *attraction*; cf. Chapters 7 and 10 (Wu, 1994).

8.4 Polar Interactions: Lewis and Brønsted Acid-Base Approaches

Following the terminology of G. N. Lewis (1923), a Lewis *acid* is any species of molecule, ion or solid that can *accept* a share in a pair of electrons during the course of a chemical reaction. Conversely, a Lewis *base* is any such species that can *donate* a share in a pair of electrons during the course of a chemical reaction.

A *subset* of Lewis acids and Lewis bases is represented by Brønsted acids and Brønsted bases. Whilst Lewis acids are electron accepters and Lewis bases are electron-donors, Brønsted acids are defined as hydrogen (or proton) *donors* and Brønsted acids are defined as hydrogen (or proton) *acceptors*. The occurrence of Brønsted acid-base interactions thus is less general than that of Lewis acid-base interactions. Because of its wider application and to avoid confusion, the Lewis terminology is the preferred one in this work.

Lewis acid-base interactions are hereafter abbreviated as AB interactions, and the superscript AB (e.g., γ^{AB}) will be used in formulas or equations as pertaining to Lewis acid-base interactions.

As this work treats colloid and surface phenomena and does not deal with covalent (or other) chemical reactions, all AB interactions treated here are restricted to the non-covalent category and remain limited to those AB interactions with a free energy ($|\Delta G^{AB}|$) of not more than 102 mJ/m^2, at 20°C. This limiting energy arises because the strongest non-covalent AB interaction one is likely to encounter in the context of colloid and surface phenomena is the hydrogen bonding free energy of cohesion between water molecules at

20°C, where ΔG^{AB}_{coh} = -102 mJ/m^2. This free energy is the driving energy for the hydrophobic attraction between apolar molecules or particles, immersed in water (van Oss and Good, 1991; van Oss 1994). We do not subscribe to Drago *et al.*'s hypothesis (1971) that electron-acceptor/electron-donor interactions ought to be subdivided into "electrostatic" and "covalent" contributions. From an electrostatic point of view, they do not fit in with electrical double layer interactions, and from a covalent point of view, they are one or more orders of magnitutde too small. In addition, the introduction of two further (unnecessary) unknowns makes practically all the applicable equations unsolvable. AB-type chemical bonds which have been argued to consist of mixtures of electrostatic and covalent bonds are therefore not considered here.

The surface-thermodynamic manifestations of AB interactions in conjunction with the interactions caused by Lifshitz-van der Waals (LW) and electrostatic (EL) forces, are discussed in Chapter 5.

8.4.1 Lewis acid-base properties of polar condensed-phase materials

All polar condensed-phase materials have *two* polar properties which both can vary independently, from one material or compound to another. These are their electron-accepticity and their electron-donicity. They are expressed as the two parameters that make up their acid-base surface tension component, γ^{AB}, i.e., their electron-acceptor parameter, $\gamma^{\oplus}$, and their electron-donor parameter, $\gamma^{\ominus}$, such that:

$$\gamma^{AB} = 2\sqrt{\gamma^{\oplus}\gamma^{\ominus}} \tag{8.1}$$

(see Chapter 5, Eq. 5.39). Even when one of these two parameters is zero (as is the rule with dry solid surfaces), that fact plays an important role in determining *whether* particles of a solid or a solute, immersed in a polar liquid, will attract or repel each other, and with what *energy* that attraction or repulsion will be endowed.

8.4.2 Polar solids

On a macroscopic level, i.e., on a level where the interactions are determined by the net outcome (e.g., stability or flocculation of a suspension, or rejection or engulfment by an advancing coagulation front, or adhesion or non-adhesion), or by contact angle measurements, *dry polar solid surfaces are monopolar.* This monopolarity is usually in favor of *electron-donicity* (van Oss *et al.*, 1997). The reason for the monopolarity of dry polar solid surfaces is the following: the cohesion between the molecules of the solid is due to their covalent as well as their LW + AB + EL interactions (LW = Lifshitz-van der Waals and EL = electrostatic; see Chapter 5). The *cohesive* bonds that are due to AB and EL forces, are no longer available for the adhesive interactions that are measurable at the surface of solids. Thus, the only surface interactions that are still measurable by macroscopic means, at the surface of a solid, are threefold:

1. *LW interactions*: atoms and molecules at a solid surface can engage in the usual Lifshitz-van der Waals interactions with other solid, solute or liquid molecules in their reasonably close vicinity.

2. *AB interactions*: after satisfying all possible cohesive electron-acceptor/ electron-donor bonds in the solid, the only AB capacity still available for reacting at the solid surface is that of the parameter that was present in the solid *in excess.* Thus only that *monopolar* capacity can be measured through its adhesive interaction with other solid, solute or liquid molecules of the opposite polarity.

3. *EL interactions*: exactly the same situation occurs as in AB interactions, only atoms or molecules with the sign of charge that still is available after all cohesive EL bonds are satisfied, will be capable of engaging in adhesive EL interactions with other solid, solute or liquid molecules of the opposite sign of charge.

Notwithstanding the undoubted presence of probably relatively small, but still non-negligible numbers of electron-acceptor sites, such as on clay particles, large crystals, glass surfaces, etc., by *macroscopic measurements* (e.g., contact angle determination), only net electron-donicity is found (van Oss *et al.*, 1997). By, e.g., protein-adsorption onto such mineral surfaces it could however be shown, that *microscopically* small, discrete electron-acceptor sites are active in the guise of plurivalent metal ions (van Oss *et al.*, 1995a,b).

8.4.3 Polar solutes

Polar polymers also tend to be monopolar when dried from aqueous solutions, or when obtained in a flat solid layer through evaporation from solution in an organic solvent. Examples are: dextran, various proteins (van Oss 1994; van Oss *et al.*, 1995b) as well as polymethyl methacrylate (van Oss 1994). Dextran is a polymer of glucose, which therefore contains many OH-groups. Nonetheless only electron-donating activity is found on dry layers of dextran, which argues for *intra* molecular electron-acceptor/electron-donor interactions. Similarly, with dried layers of low molecular weight sugars (e.g., glucose, fructose, lactose, galactose, maltose; cf. van Oss (1994)) a zero (or negligible) $\gamma^{\oplus}$ is found; here the monopolarity is mainly due to *inter* molecular electron-acceptor/electron-donor interactions (van Oss *et al.*, 1987a) that occurred during thedrying process. However, when mono- or dimeric sugars are present in aqueous solution at a fairly high concentration, they manifest strong bipolarity (Docoslis *et al.*, 2000a), as do various polar liquids (e.g., water, glycerol, formamide, ethylene glycol; see below).

The monopolarity of water-soluble polymers persists to a significant degree in aqueous solutions; this follows from their pronounced solubility as well as from the phase separation observed in mixtures of two or more polymers in aqueous solution. This is because a net repulsion is required for strong aqueous solubility of polymers, as well as for phase separation between two or more polymer species in solution, and repulsion between macromolecules and/or particles requires significant monopolarity, in the absence of strong electrical surface potentials (van Oss *et al.*, 1987a). (It should be noted that the electrical surface potential of dextran, in aqueous solution, is zero.)

The monopolarity of various types of clay particles (e.g., montmorillonite) is also the main contributing factor to their stability in water (van Oss *et al.*, 1990a; Wu *et al.*, 1994a, b; Giese *et al.*, 1996); see also Chapter 8.

8.4.4 Polar liquids

Many polar liquids are bipolar, that is, they have significant electron-donicity, as well as a non-negligible electron-acceptor. The principal exceedingly polar liquid is water, which has strong bipolarity. The polar component of its surface tension, i.e., $\gamma^{AB}_{water} = 51$ mJ/m^2, which represents 70% of its total surface tension ($\gamma_{water} = 72.8$ mJ/m^2). At 20°C it is assumed that γ^{AB}_{water} consists of equal parts of $\gamma^{\oplus}$ and $\gamma^{\ominus}$, so that then $\gamma^{\oplus}_{water} = \gamma^{\ominus}_{water} = 25.5$ mJ/m^2 (cf. Chapter 5), which makes water typically bipolar. Other bipolar liquids are the alcohols (e.g., methanol, ethanol, ethylene glycol, glycerol), and, e.g., formamide (van Oss 1994). There alsoare monopolar liquids, e.g., benzene ($\gamma^{\ominus}$), chloroform ($\gamma^{\oplus}$), methyl ethyl ketone ($\gamma^{\ominus}$) and tetrahydrofuran ($\gamma^{\ominus}$) (van Oss 1994). When solids show a non-negligible $\gamma^{\oplus}$-value (e.g., by contact angle measurements), in conjunction with a sizable $\gamma^{\ominus}$, one must immediately suspect the presence of residual liquid on the surface of the solid, which usually is residual water of hydration (van Oss *et al.*, 1997).

8.5 The Hydrophobic Effect: Hydrophobic Attraction

The hydrophobic effect is best described by Eq. 6.51, rearranged:

$$\begin{aligned}\Delta G^{IF}_{iwi} &= -\overbrace{2\gamma^{LW}_{i}}^{\text{cohesion}} - \overbrace{2\gamma^{LW}_{w}}^{\text{cohesion}} + \overbrace{4\sqrt{\gamma^{LW}_{i}\gamma^{LW}_{w}}}^{\text{adhesion}} \\ &\quad \overbrace{-4\sqrt{\gamma^{\oplus}_{i}\gamma^{\ominus}_{i}}}^{\text{cohesion}} \overbrace{-4\sqrt{\gamma^{\oplus}_{w}\gamma^{\ominus}_{w}}}^{\text{cohesion}} + \overbrace{4\sqrt{\gamma^{\oplus}_{i}\gamma^{\ominus}_{w}}}^{\text{adhesion}} + \overbrace{4\sqrt{\gamma^{\oplus}_{w}\gamma^{\ominus}_{i}}}^{\text{adhesion}}\end{aligned} \tag{8.2}$$

where the subscript, w, stands for water and the subscript, i, denotes the solute, compound or particle immersed in water. ΔG^{IF}_{iwi} is the interfacial free energy of interaction between two molecules or particles, i, in water. It can be seen that the cohesive terms are all negative (i.e., attractive), while the adhesive terms are positive (i.e., repulsive). Thus, *in toto*, attraction prevails when the total of the *cohesive* terms is larger than the total of the *adhesive* terms.

Now, *cohesion* of similar molecules or particles signifies insolubility of these molecules, or clumping of these particles. And *adhesion* between solute and liquid molecules, or between particles and the liquid, favors solubility of solute molecules, or suspension stability of particles, immersed in the liquid. Taking octane, immersed in water, as an example (see Table 8.1), the LW interaction is zero because γ^{LW}_{octane} and γ^{LW}_{w} are virtually the same. This leaves as the only term that is non-zero, $-4\sqrt{\gamma^{\oplus}_{w}\gamma^{\ominus}_{w}}$, which is the Lewis acid-base (hydrogen bonding, in this case) energy of cohesion of the water molecules, which at 20°C, is equal to -102 mJ/m^2, so that ΔG^{IF}_{iwi}, for octane = -102 mJ/m^2. This is the attractive energy between two octane molecules (or drops), immersed in water, at 20°C, and also the fundamental energy of attraction due to the hydrophobic effect. However, as can be seen from Table 8.1 and Eq. 8.2, whereas the AB cohesive energy of attraction between water molecules ($-4\sqrt{\gamma^{\oplus}_{w}\gamma^{\ominus}_{w}}$) is -102 mJ/m^2, the AB cohesive energy of attraction between octane molecules ($-4\sqrt{\gamma^{\oplus}_{i}\gamma^{\ominus}_{i}}$) is zero! Nonetheless the total free energy of attraction, ΔG^{IF}_{iwi}, in this case, equals -102 mJ/m^2, due to the AB energy of cohesion of water, and it does not matter if this AB cohesive energy originates with octane or with water, as:

$$-\frac{1}{2}\gamma_{iw} = \Delta G_{iwi} \equiv \Delta G_{wiw} \tag{8.3}$$

where, by definition, ΔG_{iwi} is completely interchangeable with ΔG_{wiw}, when both are expressed in mJ/m^2. It thus suffices for only *one* of the components in a two-component mixture to have a predominant energy of cohesion, for both of them to be insoluble, or unstable, in the other.

One important consequence of the fact that the AB cohesion energy of water (i.e., -102 mJ/m^2 at 20°C) is the underlying cause of the hydrophobic effect, is that *this attractive energy is always present in all interactions in aqueous media.* Strongly hydrophilic solutes or solids must therefore furnish a repulsive energy that is greater than the 102 mJ/m^2 of the AB cohesion of water *plus* the (usually much smaller) van der Waals attraction. This is illustrated in Table 8.1. Clearly, the AB energy of cohesion of water is always there (column C). At the top of Table 8.1 are the most hydrophobic compounds or particles and at the bottom are the most hydrophilic ones. It should be noted that dry solids have $\gamma^{\oplus}$-values that are exceedingly small or zero (van Oss *et al.*, 1997). Muscovite and glass are extremely hydrophilic, which entails that their surfaces always are somewhat hydrated, when exposed to ordinary humid air, hence their $\gamma^{\oplus}$-values of 1.8 and 1.3 mJ/m^2, respectively. This does contribute somewhat to the repulsion (column D of Table 8.1), but is more than compensated by the AB cohesion at the surface of the material itself (column B). Other factors being equal, the strongest repulsion always occurs between monopolar surfaces, which in these two cases would be in the absence of $\gamma^{\oplus}$- values. It can also be seen in Table 8.1 that very hydrophilic materials (muscovite, glass and PEO) whose molecules or particles repel each other when immersed in water, must have a high $\gamma^{\ominus}$-value (typically well over 28.5 mJ/m^2) to produce an interaction with the $\gamma^{\oplus}_{w}$ of water, that is significantly higher than the attractive hydrophobic effect energy of 102 mJ/m^2.

It is not always realized that hydrophobic interactions can quite readily take place between one hydrophobic and one hydrophilic molecule or particle, immersed in water, see Eq. 5.50. This is demonstrated in Table 8.2. Here the LW, the AB and the total interfacial (IF) free energies are shown of the interaction of polyethylene (PEO, which is one of the most hydrophilic materials known) with hydrophobic, mildly hydrophobic and hydrophilic entities. It clearly shows that, in water, PEO will bind to the more hydrophilic substrata (i.e., Teflon, octane, talc). PEO will not bind (in water) to the only slightly hydrophobic smectite, hectorite, and it is even more strongly repelled (on a macroscopic scale) by the very hydrophilic surfaces of muscovite and glass. It should be noted that these considerations only apply to interactions on a macroscopic level. Even when a macroscopic repulsion exists on a macro-

Material (values in mJ/m^2)	$-2(\sqrt{\gamma_i^{LW}} - \sqrt{\gamma_w^{LW}})^2$ A	$-4(\sqrt{\gamma_i^{\oplus}\gamma_i^{\ominus}}$ B	$+\sqrt{\gamma_w^{\oplus}\gamma_w^{\ominus}}$ C	$-\sqrt{\gamma_i^{\oplus}\gamma_w^{\ominus}}$ D	$-\sqrt{\gamma_w^{\oplus}\gamma_i^{\ominus}})$ E	$= \Delta G_{iwi}^{IF}$
Teflon FEP $\gamma^{LW}=17.9; \gamma^{\oplus} = \gamma^{\ominus} = 0$	-0.4	0	-102.0	0	0	-102.4
octane $\gamma^{LW} = 21.6; \gamma^{\oplus} = \gamma^{\ominus} = 0$	0	0	-102.0	0	0	-102.0
talc (dry) $\gamma^{LW} = 34.2; \gamma^{\oplus} = 0.2; \gamma^{\ominus} = 6.9$	-2.8	-3.3	-102.0	+6.4	+53.0	-48.7
hectorite (dry) $\gamma^{LW} = 39.9; \gamma^{\oplus} = 0; \gamma^{\ominus} = 23.7$	-5.4	0	-102.0	0	+98.3	-9.1
muscovite $\gamma^{LW} = 40.6; \gamma^{\oplus} = 1.8; \gamma^{\ominus} = 51.5$	-5.8	-38.5	-102.0	+27.0	+145.0	+25.7
clean glass slide $\gamma^{LW} = 33.7; \gamma^{\oplus} = 1.3; \gamma^{\ominus} = 62.2$	-2.6	-36.0	-102.0	+23.0	+159.3	+41.7
polyethylene (PEO-6000) $\gamma^{LW} = 43.0; \gamma^{\oplus} = 0; \gamma^{\ominus} = 64$	-7.1	0	-102.0	0	+161.6	+52.5

Table 8.1: *Components of the interfacial free energy of interaction between molecules or particles of material, i, immersed in water, w, in mJ/m^2, at 20°C. Component* C *represents the hydrophobic interaction energy, or hydrophobic effect energy; sensu stricto, see eq. 8.2.*

scopic level, on a microscopic scale an attraction can exist at various small sites, which when one or both entities are appropriately oriented toward each other, can prevail locally and give rise to a net attraction (van Oss *et al.*, 1995a); see also Sections 9.7.3 and 9.7.4. It should also be stressed that it is possible to have a *repulsive* Lifshitz-van der Waals interaction, between two *different* entities which are immersed in a third liquid, cf. Eq. 6.32, which can also be written as:

$$\Delta G^{LW}_{132} = 2\left(\sqrt{\gamma_1^{LW}} - \sqrt{\gamma_3^{LW}}\right)\left(\sqrt{\gamma_2^{LW}} - \sqrt{\gamma_3^{LW}}\right) \tag{6.32a}$$

from which it can be seen that ΔG^{LW}_{132} is positive (i.e., repulsive) when the value of γ_3^{LW} (of the liquid) lies in between the values of γ_1^{LW}, and γ_2^{LW} (Visser, 1972; van Oss 1994). In Table 8.2 it can be seen that the PEO-Teflon interaction in water is a clear case of $\Delta G^{LW}_{1w2} > 0$; the van der Waals repulsion, in water, in this case of course is overwhelmed by the large hydrophobic attraction ($\Delta G^{AB}_{1w2} < 0$). As a consequence of the strong underlying hydrophobic effect that is peculiar to the properties of water, one rarely encounters a net repulsion between two different entities immersed in water, that is entirely due to repulsive van der Waals forces. In apolar liquids, or in some monopolar liquids, however, repulsion solely due to LW interactions are quite common, cf. van Oss *et al.* (1989a); van Oss (1994).

Largely as a consequence of the fact that in the past most experimental work on the hydrophobic effect has been done with alkanes (see, e.g., Tanford (1980)), the myth has arisen that hydrophobic interactions are mainly entropic (see, e.g., Hiemenz (1986); Israelachvili (1992)). In the case of alkanes this is indeed the case. However, when one actually measures the values of ΔG^{IF}_{iwi} at different temperatures, using van't Hoff's equation:

$$\Delta H = -\frac{\delta \Delta H}{\delta T_P} \tag{8.4}$$

in conjunction with:

$$\Delta G = \Delta H - T\Delta S \tag{8.5}$$

both the enthalpy, ΔH (Eq. 8.4) and the entropy, S (Eq. 8.5) can be found. For instance, van Oss and Good (1991) and van Oss (1994) found that ΔG^{IF}_{iwi} for alkanes (e.g., hexane) is indeed mainly entropic, but for CCl_4, dibromoethane and heptanoic acid, ΔG^{IF}_{iwi} is mainly enthalpic, and for benzene it is about equally entropic and enthalpic, at ≈20°C. In all cases of interaction between apolar, or slightly polar molecules or particles interacting in water,

PEO interacting with	ΔG^{LW}_{1w2}	ΔG^{AB}_{1w2}	$= \Delta G^{IF}_{1w2}$
Teflon	1.66	-21.21	-19.55
octane	0.08	-21.21	-21.13
talc	-4.45	+3.45	-1.0
hectorite	-6.22	+27.96	+21.74
muscovite	-6.43	+43.36	+36.93
glass	-4.29	+51.72	+47.43

Table 8.2: *Lifshitz-van der Waals and Lewis acid-base interaction energies, ΔG_{1w2} (cf. Eq. 6.50) between poly (ethylene oxide)(PEO-6000), dissolved in water, with hydrophobic and hydrophilic compounds (see Table 8.1), at 20°C, in mJ/m^2.*

the hydrophobic attraction energy is likely to be predominant, in comparison with the Lifshitz-van der Waals attraction and the electrostatic repulsion energies.

8.6 Hydrophilic Repulsion

As is clear from the three hydrophilic items shown at the bottom of Table 8.1, (and also from the other items shown in that table), by its very nature, the strong attractive free energy of the hydrophobic effect is always present in all interactions taking place in water. What makes a compound or particle hydrophilic (i.e., repulsive in water), is its ability to achieve a Lewis acid-base repulsion that is significantly larger than the underlying hydrophobic attraction. (It should be remembered that usually a fairly small, but non-negligible Lifshitz-van der Waals attraction must also be surmounted, cf. Tables 8.1 and 8.2.

The strong hydrophilic repulsion that is generated by very hydrophilic molecules and particles most often originates from their dominant monopolar electron-donicity (i.e., from their low to zero $\gamma_i^{\oplus}$ and their high $\gamma_i^{\ominus}$-value, cf. term E of the ΔG^{IF}_{iwi} equation at the top of Table 8.1), due to exposed oxygen at their surface. The predominance of this oxygen-caused electron-donicity, in the form of oxides and hydroxides in its turn is a consequence of the prevalence of oxygen in the Earth's lower atmosphere (about 20 vol%) and in the Earth's surface (about 46 wt%) (van Oss *et al.*, 1997).

Hydrophilic repulsion, due to monopolar electron-donicity (van Oss *et al.*, 1987a) is the principal cause of the aqueous solubility of biopolymers, of the stability of circulating bloodcells and of the aqueous suspension stability of alumina (corundum), silica (quartz), kaolinite, montmorillonite, a synthetic zeolite, vermiculite, and other mineral particles, cf. Tables 9.1 and 9.2.

8.7 Definition of Hydrophobicity and Hydrophilicity

When a substance or surface is called "hydrophobic", this implies that it fears water, a designation which is however incorrect. Even the most "hydrophobic" or, more correctly, the most apolar of substances binds water with a considerable amount of energy. For example, Teflon ($\gamma_i^{LW} = 17.9$; $\gamma_i^{\oplus} = \gamma_i^{\ominus} = 0$ mJ/m^2), binds water with a free energy, $\Delta G_{iw} = -39.5$ mJ/m^2. The most hydrophilic of materials, e.g., polyethylene oxide (PEO) ($\gamma_i^{LW} = 43.0$, $\gamma_i^{\oplus} = 0$, $\gamma_i^{\ominus} = 64$ mJ/m^2), binds water with a free energy, $\Delta G_{iw} = -142$ mJ/m^2, which is about 3.6 times stronger than for Teflon, but Teflon's hydration energy still is far from negligible. It is possible to make an empirical scale of hydrophobicity/hydrophilicity, based on the material's hydration energy, whereby materials with a ΔG_{iw}-value that is more negative than about -113 mJ/m^2 are hydrophilic, and materials for which ΔG_{iw} is less negative than approximately -104 to -116 mJ/m^2, are hydrophobic (van Oss 1994); see Table 8.3.It should be emphasized that the free energy of hydration *cannot be measured* by calorimetry: that method only yields a value for the enthalpy, ΔH_{iw} which, in the absence of information about the entropic contribution ($T\Delta S$; cf. Eq. 8.5) to ΔG_{iw}^{IF}, does not allow one to arrive at any conclusion at all concerning ΔG_{iw} (contrary to, e.g., Low (1961)). ΔG_{iw} is best derived from the measured values for γ_{iw}, γ_i, and γ_w and then using Dupré's equation (Eqs. 5.29 and 5.42).

Another yardstick for hydrophobicity or hydrophilicity is the value of $\gamma_i^{\ominus}$: materials which have a γ_i^{LW}-value of about 40 mJ/m^2, with a $\gamma_i^{\ominus}$-value higher than 28.4 mJ/m^2 are hydrophilic and at $\gamma_i^{\ominus} < 28.4$ mJ/m^2 they are hydrophobic; see also Table 8.3.

The most rational, absolute scale of hydrophobicity/hydrophilicity however is the one defined by the interfacial free energy of interaction of molecules, particles or surfaces of material, i, with one another, when immersed in water, i.e., ΔG_{iwi}^{IF}. Quite simply, when $\Delta G_{iwi}^{IF} < 0$, i is hydrophobic and when $\Delta G_{iwi}^{IF} > 0$, i is hydrophilic. The degree of hydrophobicity is indicated by the amount with which $\Delta G_{iwi}^{IF} < 0$, and the degree of hydrophilicity by the

Material (γ_i in mJ/m^2)	ΔG^{IF}_{iwi}	ΔG_{iw}	$\gamma^{\ominus}_i$
Hydrophobic			
Teflon	-102.4	-39.5	0
$\gamma_i^{LW} = 17.9$; $\gamma_i^{\oplus} = \gamma_i^{\ominus} = 0$			
talc (dry)	-45.9	-84.3	6.9
$\gamma_i^{LW} = 34.2$; $\gamma_i^{\oplus} = 0.1$; $\gamma_i^{\ominus} = 6.9$			
hectorite	-9.1	-108.2	23.7
$\gamma_i^{LW} = 39.9$; $\gamma_i^{\oplus} = 0$; $\gamma_i^{\ominus} = 23.7$			
Hydrophilic			
palygorskite	+5.0	-104.8	28.7
$\gamma_i^{LW} = 29.5$; $\gamma_i^{\oplus} = 0$; $\gamma_i^{\ominus} = 28.7$			
clean glass slide	+41.7	-136.4	62.2
$\gamma_i^{LW} = 33.7$; $\gamma_i^{\oplus} = 1.3$; $\gamma_i^{\ominus} = 62.2$			
PEO-6000	+52.5	-142.0	64.0
$\gamma_i^{LW} = 43.0$; $\gamma_i^{\oplus} = 0$; $\gamma_i^{\ominus} = +64.0$			

Table 8.3: *Three criteria for distinguishing hydrophobic from hydrophilic materials: the defining value, ΔG^{IF}_{iwi}; the hydration energy, ΔG_{iw}; the electron-donor parameter, $\gamma^{\ominus}$.*

amount with which $\Delta G^{IF}_{iwi} > 0$. At $\Delta G^{IF}_{iwi} = 0$, the material is neutral, i.e., it is then neither hydrophilic nor hydrophobic (van Oss and Giese,1995). Table 8.3 gives a comparison between the absolute measure (ΔG^{IF}_{iwi}) and ΔG_{iw} and $\gamma^{\ominus}_{iw}$, as criteria for hydrophobicity/hydrophilicity of various compounds and materials. (Electrostatic interactions, in the guise of ΔG^{EL}_{iwi}, can also be of some influence on the apparent hydrophobicity or hydrophilicity of polymers or particles suspended in water, but as ΔG^{EL}_{iwi} depends on the ζ-potential, which is measured separately and which is strongly dependent upon the ionic strength of the aqueous medium, it cannot, for practical reasons, be considered part of ΔG^{IF}_{iwi}, i.e., part of the hydrophobicity/hydrophilicity scale).

Table 8.3 shows a few examples of hydrophobic and hydrophilic entities. The defining value, ΔG^{IF}_{iwi}, is clearly the most foolproof: for neutral particles its sign indicates whether instability ($\Delta G^{IF}_{iwi} < 0$) or stability ($\Delta G^{IF}_{iwi} > 0$) of an aqueous suspension is favored, and in either case the value of $|\Delta G^{IF}_{iwi}|$ indicates the degree of instability or stability. The same holds true for the aqueous solubility of solutes (cf. Section 9.6).

The free energy of hydration, ΔG_{iw}, gives one a semi-quantitative idea of hydrophobicity (ΔG_{iw} less negative than about -104 to -116 mJ/m^2) or hydrophilicity (ΔG_{ie} more negative than the above values), but particularly when ΔG_{iw} is close to these cut-off values, nothing can be said with certainty; cf. ΔG_{iw} = -108.2 mJ/m^2 for the slightly hydrophobic smectite, hectorite, and ΔG_{iw} = -104.2 for palygorskite, which is hydrophilic. The value of $\gamma^{\ominus}_{i}$ is also semi-quantitatively indicative of hydrophobicity or hydrophobicity, but here again, when close to the cut-off of, say $\gamma^{\ominus}_{i} \approx 28.4$ mJ/m^2, uncertainty reigns. When $\gamma^{LW}_{i} > 42$ mJ/m^2, i would still be slightly hydrophobic, but when $\gamma^{LW}_{i} \leqslant 40$ mJ/m^2, i would be hydrophilic. Incomplete dryness, which gives rise to a non-negligible $\gamma^{\oplus}_{i}$-value, further increases the uncertainty of the cut-off.

Thus, the sign and value of the free energy of the interfacial interaction between molecules, particles, or surfaces, i, immersed in water (ΔG^{IF}_{iwi}) are the only really reliable quantitative criteria for defining hydrophobicity or hydrophilicity (van Oss and Giese, 1995).

8.8 DLVO Approach, Including Lewis Acid-Base Energies

Particle stability, like solute solubility, is favored when, at contact, $\Delta G^{TOT}_{iwi} >$ 0 (see Eq. 8.6, below). However, in the case of particles with radii R $\gtrsim$ 10 to 100 nm, a certain measure of stability may occur, even if, at contact (i.e., at $\ell = \ell_0$), $\Delta G^{\mathrm{TOT}}_{\mathrm{iwi}} < 0$. This is due to the fact that at certain interparticle distances, $\ell > \ell_0$, it may well occur that $\Delta G^{\mathrm{TOT}}_{\mathrm{iwi}}(\ell) > 0$. To ascertain the values which $\Delta G^{\mathrm{TOT}}_{\mathrm{iwi}}$ can assume, as a function of the interparticle distance, ℓ, in a given particle suspension, $\Delta G^{\mathrm{TOT}}_{\mathrm{iwi}}$ must be plotted vs. ℓ, in a given particle suspension, to which effect ΔG_{iwi} must be separately plotted vs. ℓ for all three different functions (LW, AB, EL), and then combined into a $\Delta G^{\mathrm{TOT}}_{\mathrm{iwi}}$ vs. ℓ plot.

This approach (comprising, however, only the LW and EL forces) was developed independently by Derjaguin and Landau (1941) and Verwey and Overbeek (Verwey and Overbeek, 1948, 1999), and became known, after these authors, as the DLVO theory, and the corresponding energy vs. distance plots as DLVO plots. In the absence, or virtual absence, of polar (AB) interactions (i.e., in mainly apolar media), the DLVO theory correlates admirably with the stability of particle suspensions. However, in the cases of particle suspensions in polar and especially aqueous media, disregarding the influence of polar interactions by using simple DLVO plots usually leads to severely unrealistic models (van Oss *et al.*, 1990a). In other words, in all polar systems one should take into account ΔG_{131}, as a function of ℓ, as:

$$\Delta G^{\mathrm{TOT}}_{131}(\ell) = \Delta G^{\mathrm{LW}}_{131}(\ell) + \Delta G^{\mathrm{EL}}_{131}(\ell) + \Delta G^{\mathrm{AB}}_{131}(\ell) \tag{8.6}$$

8.8.1 Decay of interaction energies and forces as a function of distance

Table 8.4 shows the functions of free energy (ΔG) and force (F) vs. distance (ℓ) for flat parallel plates, a sphere and a flat plate, and two spheres of equal radius, R, for LW, EL and AB energies and forces (see also Section 5.12.)

It should be noted that for each of the three different modes of interaction (LW, AB and EL), for the interactions involving one or more spherical bodies or molecules, both the interaction energies and the interaction forces *are proportional to the spheres' radii*, R. Thus, approximately spherical particles with irregular protrusions on their surfaces, with a smaller radius of curvature, r, will locally undergo a much smaller repulsion than smooth particles with a radius R, so that particles with protrusions of radius, r, can approach each other much more closely than smooth particles. For irregularly shaped particles such as clay particles, this is an extremely important consideration.

8.8.2 The extended DLVO (XDLVO) approach applied to aqueous media

In water, polar (AB) interactions, whether attractive or repulsive, are practically never negligible. Either way, AB interactions in aqueous media usually are, at close range, one or two orders of magnitude greater than LW or EL interactions. Some fairly typical examples of this are shown in Figures 8.1 and 8.4. Here, for the sake of uniformity, rather commonly used latex particles with a diameter, R, of approximately 0.4 μm are chosen as an example.

In the hydrophilic case (Figures 8.1 and 8.4) they are taken to be coated with a hydrophilic material such as hydrated human serum albumin (see Table 8.4). Human serum albumin is among the most highly (negatively) charged plasma proteins; its ψ_o-potential of $\approx$-30 mV is therefore taken as a typical high value for a proteinaceous material. The Hamaker constant of hydrated albumin with respect to water (A_{iwi}) is quite low; it is only of the order of 5×10^{-22} J, due to the close resemblance between the LW properties of the outer layer of *hydrated* protein and those of the aqueous medium. The polar repulsion between hydrated albumin molecules is rather strong; ΔG^{AB}_{iwi} is of the order of $+20$ mJ/m^2 at close range. The decay length of water is here taken to be $\lambda \approx 0.6$ nm (van Oss 1994). The energy balances are shown at high salt concentrations (0.1 M NaCl), which is close to the salt concentration of physiological saline solutions (Figure 8.1), as well as at low ionic strength (0.001 M NaCl) (Figure 8.4). In both cases the suspensions are stable.

At high ionic strength there is a slight secondary minimum of attraction at about 8.5 nm, but that attraction amounts to less than 1 kT. However, for larger particles or cells of comparable properties, the attraction at the secondary minimum could easily amount to, e.g., 10 kT, as the free energy of attraction (or repulsion) is proportional to the radius, R (see above). For some particles, a weak attraction at the secondary minimum may be of some importance. For single protein molecules however, with a radius R that is, typically, 100 $\times$ smaller than 0.4 μm(i.e., the radius of the particles in Figures 8.1-8.4), attraction at the secondary minimum is negligible.

8.8.3 Stability versus flocculation of aqueous particle suspensions

Complete stability of particle suspensions in liquids can only exist when ΔG_{1w1}, at $\ell = \ell_o$, has a positive value, i.e., when a sizable net repulsion prevails at close range (cf. Figures 8.1 and 8.2). In practice, to achieve solid

A. Unretarded Lifshitz-van der Waals interactions

Configuration	ΔG_ℓ^{LW}	F_ℓ^{LW}
semi-infinite parallel slabs	$\frac{-A}{12\pi\ell^2}$	$\frac{A}{12\pi\ell^3}$
sphere of radius R and semi-infinite flat slab; also valid for two crossed cylinders at 90°	$\frac{-AR}{6\ell}$	$\frac{AR}{6\ell^2}$
two spheres of radius R	$\frac{-AR}{12\ell}$	$\frac{AR}{12\ell^2}$

B. Electron donor-acceptor interactions

Configuration	ΔG_ℓ^{AB}	F_ℓ^{AB}
flat parallel plates	$\Delta G_{\ell_0}^{AB"} \exp[(\ell_0 - \ell)/\lambda]$	$F_{\ell_0}^{AB"} \exp[\ell_0 - \ell/\lambda]$ $= - \Delta G_{\ell_0}^{AB"} (1/\lambda) \exp[(\ell_0 - \ell)/\lambda]$
sphere of radius R and flat plate; also valid for two crossed cylinders at 90°	$\pi R\lambda\Delta G_{\ell_0}^{AB"} \exp[(\ell_0 - \ell)/\lambda]$	$-2\pi R\Delta G^{AB"} \exp[(\ell_0 - \ell)/\lambda]$
two spheres of radius R	$\pi R\lambda\Delta G_{\ell_0}^{AB"} \exp[(\ell_0 - \ell)\lambda]$	$-\pi R\Delta G_{\ell_0}^{AB"} \exp[(\ell_0 - \ell)\lambda]$

C. Electrostatic interactions

Configuration	ΔG_ℓ^{EL}	F_ℓ^{EL}
flat parallel plates	$(1/\kappa)64nkT\gamma_0^2 \exp(-\kappa\ell)$	$-64nkT\gamma_0^2 \exp(-\kappa\ell)$
sphere of radius R and flat plate; also valid for two crossed cylinders at 90°	$\epsilon R\psi_0^2 \ln[1 + \exp(-\kappa\ell)]$	$\epsilon\kappa R\psi_0^2 \ln[1 + \exp(-\kappa\ell)]$
two spheres of radius R	$\frac{1}{2}\epsilon R\psi_0^2 \ln[1 + \exp(-\kappa\ell)]$	$-\frac{1}{2}\kappa\epsilon R\psi_0^2 \ln[1 + \exp(-\kappa\ell)]$

Table 8.4: *The energies of interaction (ΔG_ℓ) and forces (F_ℓ) as a function of inter-surface distance, ℓ. For explanations of the symbols, see the text. Note that* $\gamma_0 = [\exp(ve\psi_0/2kT) - 1]/[\exp(ve\psi_0/2kT) + 1]$.

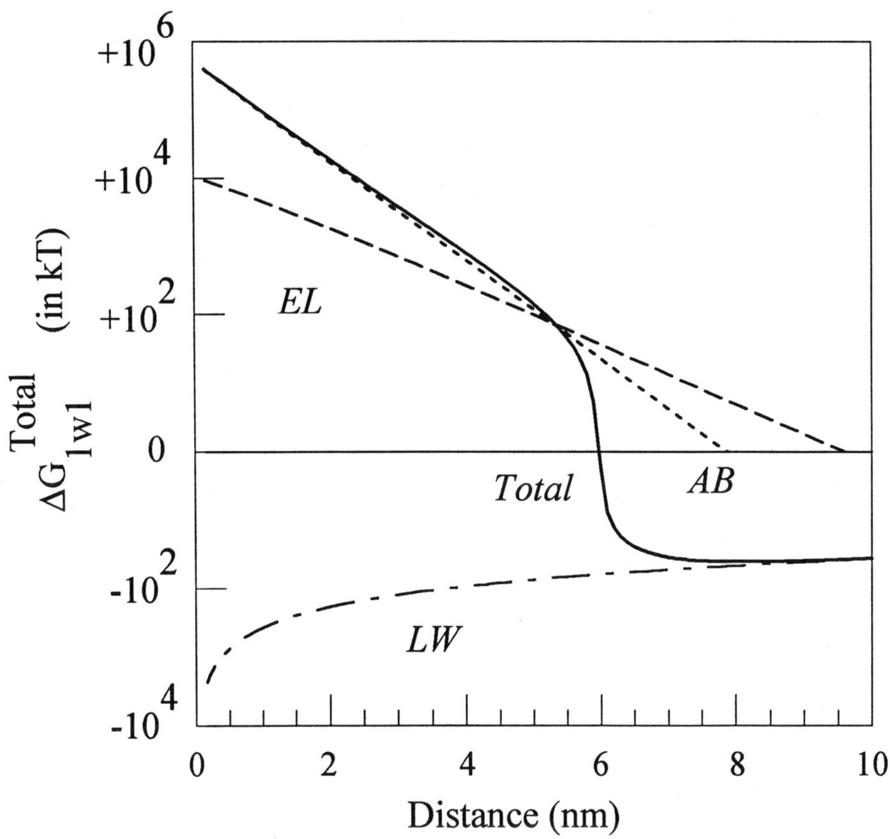

Figure 8.1: *Energy balance (ΔG_{1w1} and its components, expressed in kT) of typical spherical hydrophilic particles at neutral pH in the presence of 0.1 M NaCl: particle radius = 0.4 μm, ζ-potential = -16.3 mV, γ^{LW} = 26.6 mJ/m^2, $\gamma^{\oplus}$ = 6.3 mJ/m^2, $\gamma^{\ominus}$ = 50.6 mJ/m^2, λ = 0.6 nm, z = 0.3 nm;* $1/\kappa$ *= 1.0 nm.*

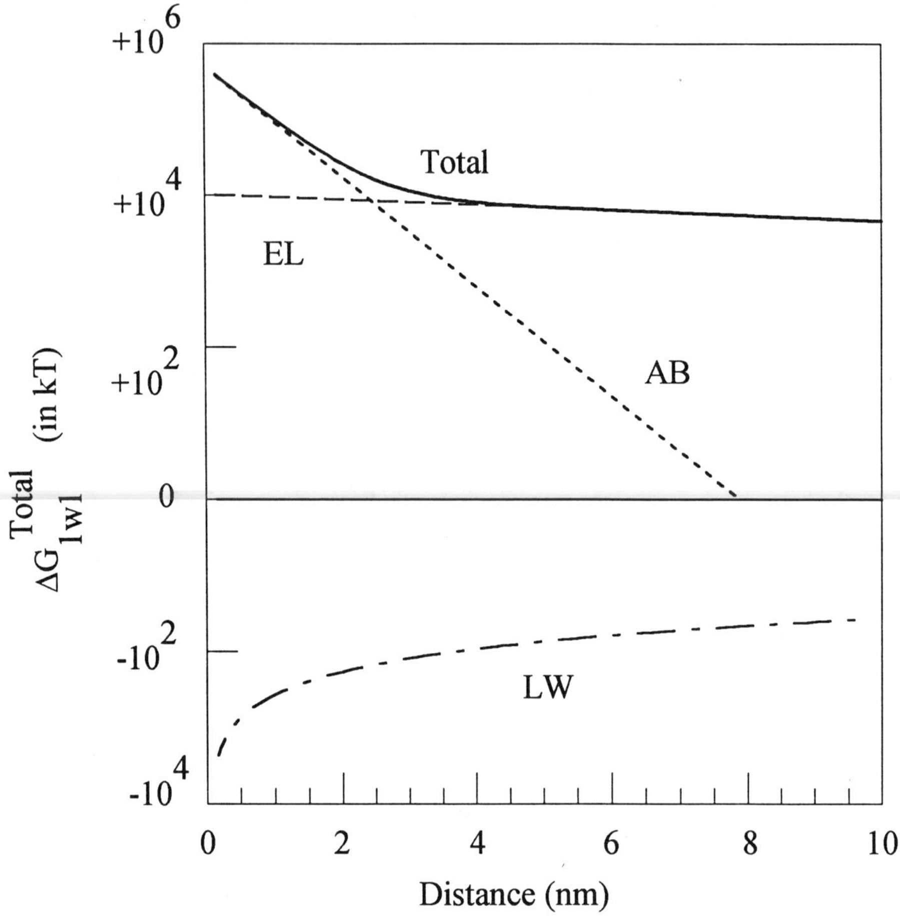

Figure 8.2: *Energy balance (ΔG_{1w1}^{Tot} and its components, expressed in kT) of typical spherical hydrophilic particles at neutral pH in the presence of 0.001 M NaCl: particle radius = 0.4 μm, ζ-potential = -16.3 mV, γ^{LW} = 26.6 mJ/m², $\gamma^{\oplus}$ = 6.3 mJ/m², $\gamma^{\ominus}$ = 50.6 mJ/m², λ = 0.6 nm, z = 0.3 nm; 1/κ = 10 nm.*

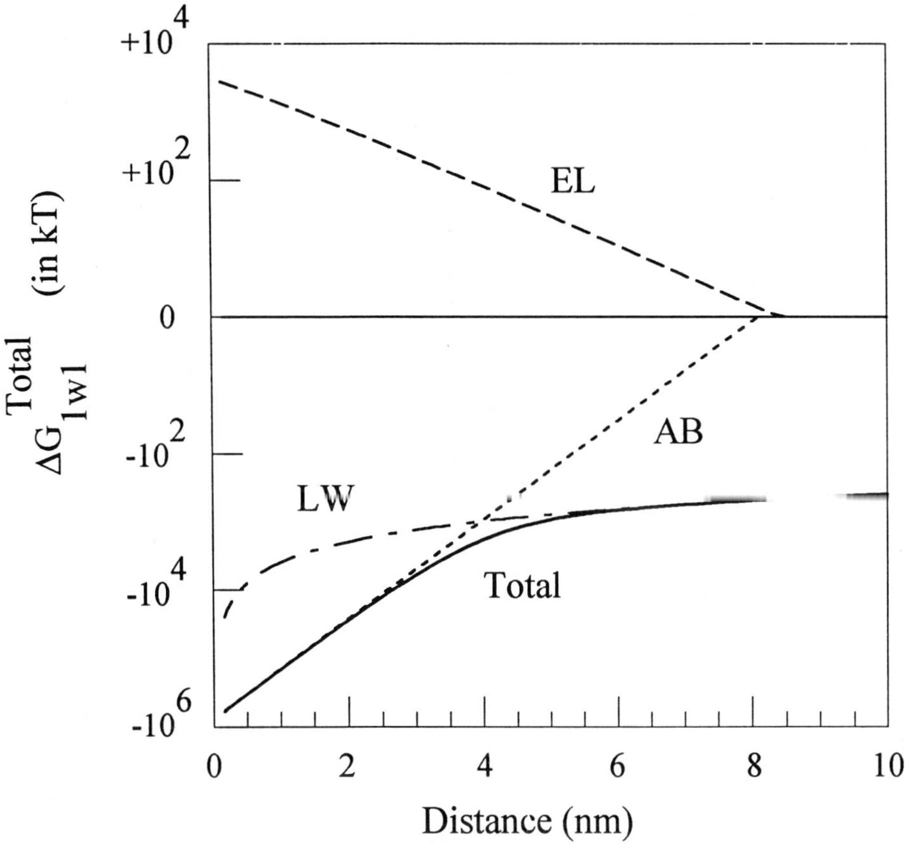

Figure 8.3: *Energy balance ($\Delta G^{\text{Total}}_{1w1}$ and its components, expressed in kT) of typical spherical hydrophobic particles at neutral pH in the presence of 0.1 M NaCl: particle radius = 0.4 μm, ζ-potential = -9 mV, γ^{LW} = 39.1 mJ/m², $\gamma^{\oplus}$ = 0.3 mJ/m², $\gamma^{\ominus}$ = 11.5 mJ/m², λ = 0.6 nm, z = 0.3 nm; 1/κ = 1.0 nm.*

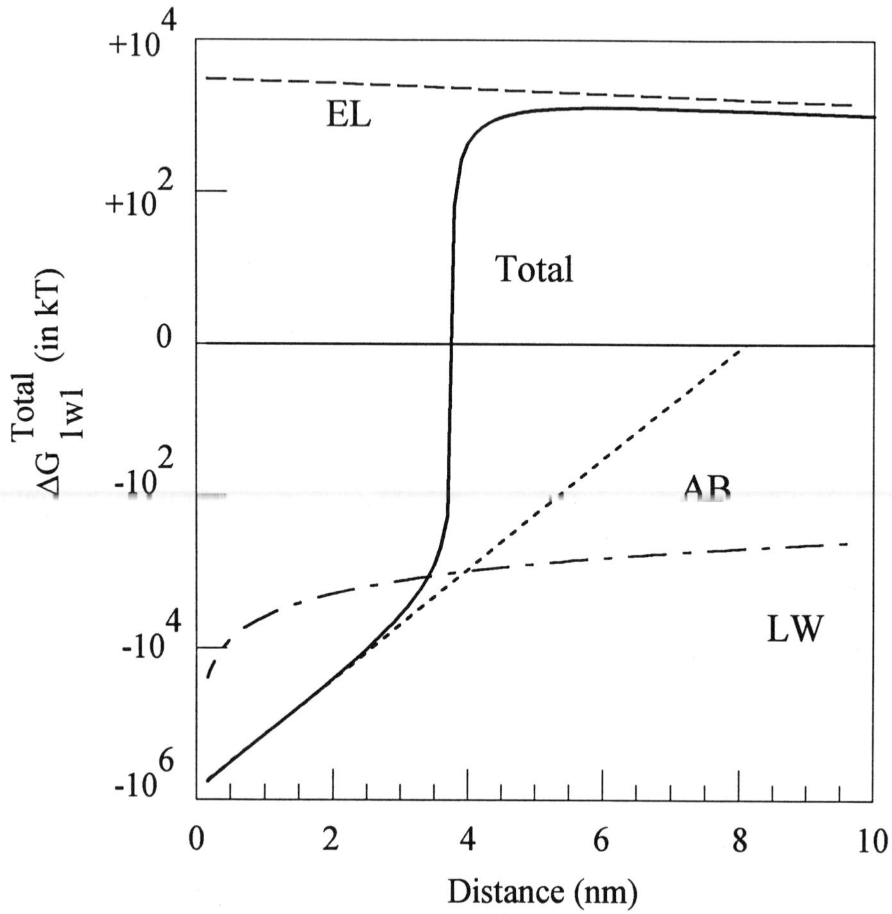

Figure 8.4: *Energy balance ($\Delta G^{\text{Total}}_{1w1}$ and its components, expressed in kT) of typical spherical hydrophobic particles at neutral pH in the presence of 0.001 M NaCl: particle radius = 0.4 μm, ζ-potential = -9 mV, γ^{LW} = 39.1 mJ/m², $\gamma^{\oplus}$ = 0.3 mJ/m², $\gamma^{\ominus}$ = 11.5 mJ/m², λ = 0.6 nm, z = 0.3 nm;* $1/\kappa$ = *10 nm.*

stability in such cases, it is necessary for $\Delta G \gg +10$ kT, at $\ell = \ell_o$. The existence of a secondary minimum of attraction at $\ell > \ell_o$ (cf. Figure 8.1) does not cause instability or flocculation. However, in the presence of large asymmetrical polymer molecules, dissolved in the liquid medium in which the particles are suspended through cross-binding by the polymer molecules, a certain degree of loosely bound particle complexes may occur, which themselves remain in stable suspension. Such particle complexes tend to dissociate again into single dispersed particles upon shaking. Complexes of this type do not give rise to flocculation, but they do have a tendency to *sediment* more quickly than single particles. One well-known example of this type of complex formation due to attraction at the secondary minimum, is the formation of rouleaux of red cells, suspended in media (e.g., plasma) containing increased amounts of large asymmetrical polymers (e.g., immunoglobulin-M, or fibrinogen) (van Oss 1994). Such rouleaux are mainly formed *in vivo* when there is an abnormally high concentration of such asymmetrical proteins in a patient's plasma, which gives rise to an increased sedimentation rate of the blood of the patient, which is used as an indication of the possible existence of an inflammation. Rouleaux of this type are clearly visible under the microscope.

Temporary stability (metastability) can occur even when there is a strong attraction between particles at short range ($\ell = \ell_o$), if there is a secondary maximum of repulsion (usually occurring at low ionic strengths). This is illustrated in Figure 8.4, where a secondary maximum of repulsion of about +10 kT is depicted, at an interparticle distance of about 6 nm. The reason for the rather frequent occurrence of metastability at low ionic strengths (i.e., in the extremely gradual decay of ΔG^{EL}_{1w1}, due to the concomitantly high value of $1/\kappa$, i.e., $1/\kappa \gg \lambda$ (cf. Table 8.1). For instance, at an ionic strength of 0.001, $1/\kappa = 10.0$ nm, corresponding to $\ln[1 + \exp(-\kappa/\ell)] = 0.31$, at $\ell = 10$ nm, so that an initial EL repulsion (at $\ell = \ell_o$) of about +500 kT still has a value of about +157 kT at 10 nm separation. Thus, a long-range secondary maximum peak of repulsion persists because (e.g., with hydrophobic particles), the negative value of ΔG^{AB}_{1w1} decays much more sharply, as it is not sensitive to changes in ionic strength at sub-molar salt concentrations of 1:1 electrolytes, and thus is only a function of $\exp(\ell/\lambda)$, where λ remains at a constant 0.6 nm. At high ionic strengths, with hydrophobic particles, no secondary maximum of repulsion can occur, because under those conditions, $1/\kappa \leq \lambda$; cf. Figure 8.3.

In general, with negatively charged particles [as is the case with most clay particles, cf. Giese *et al.* (1996)], the lower the ionic strength, the greater the stability. This is because at low ionic strength, the thickness of the diffuse ionic double layer $1/\kappa$, is at its maximum, which gives rise to a maximum EL repulsion (cf. Table 8.4-C). However, a curious but important exception occurs with kaolinite. Suspensions of kaolinite particles are stable at high ionic strengths and become unstable at low ionic strengths (Schofield and Samson, 1954). This is because kaolinite does not depend on maximum repulsion to achieve stability, but needs a minimum of *attraction* to form the most stable suspension, on account of the fact that kaolinite (while negatively charged on average), actually has a strong negative charge on its platey surfaces, but also a non-negligible positive charge at the edges of the plates. [Kaolinite shows a significantly higher negative ζ-potential in phosphate buffers than in a non-buffered 1-1 electrolyte of the same ionic strength (Giese *et al.*, 1996)]. Thus, kaolinite particles tend to combine by electrostatic *attraction* between oppositely charged edges and plates, thus flocculating in a "house-of-cards" configuration (van Olphen, 1977), which is favored at low ionic strength, when plus-minus EL attraction is at a maximum. At high ionic strength this plus-minus EL attraction is at a minimum, so that then a stable suspension is favored, see Section 8.10.

8.8.4 Inadequacy of "steric" stabilization theories

There is a school of thought which has it that hydrophobic particles that have neutral water-soluble linear polymer molecules attached to them, form stable suspensions in water, because of some "steric" activity that has been imparted to them by these polymers; see, e.g., Napper (1983). This explanation does not, upon further analysis, appear very cogent, and is not open to significant quantitative treatment. It served, however, for some time, in a vague and qualitative fashion, to illustrate if not to explain, how nonionic hydrophilic oligomers or polymers could impart stability to otherwise unstable particles in aqueous suspensions. The need for some such explanatory scenario disappeared however by the late 1980's, when the effects and the mechanism of net hydrophilic repulsion forces were elucidated (see, e.g., van Oss *et al.* (1987a, 1988a)). The actual effect of hydrophilic nonionic surfactants, oligomers or polymers, attached to hydrophobic particles, is to impart a net hydrophilic repulsion to them, when immersed in water. Perusal of Napper's work (1983) shows that "sterically" stabilizing polymers, active in aqueous suspensions, are all exceedingly water-soluble and indeed consist, in 99% of the cases, of polyethylene oxide and, in some instances, of polyvinyl alcohol. These polymers, when attached to hydrophobic particles cause them to repel one another when immersed in water, because the polymers themselves mutually repel each other, in water, through repulsive hydrophilic interactions, being exceedingly strong monopolar electron donors (van Oss 1994).

8.8.5 The extended DLVO approach in aqueous media; comparison with experimental data

Whilst the classical DLVO theory (cf. Section 8.8) only considers the interplay between attractive van der Waals and repulsive electrostatic forces, in interactions in aqueous media LW and EL forces *together* usually do not represent more than about 10% of the total non-covalent interactions. This is because, in water, all non-aqueous molecules and particles are exposed to, *inter alia*, the relatively tremendous Lewis acid-base free energy of cohesion of the water molecules, which accounts for about 70% of the cohesion of water. This is the source of the hydrophobic effect (cf. Section 8.5), *which is always present.* And in the cases where the hydrophobic attractive energy (due to the AB cohesion of water) can be surmounted, a repulsion occurs which originates from the strong AB interaction between the electron-donors of hydrophilic molecules or particles and the electron-acceptor moieties of the surrounding water molecules. Thus, neglecting AB interactions in energy versus distance diagrams in aqueous media, leads to theoretical predictions concerning, e.g., particle stabilities, which can strongly contrast with experimental reality.

8.8.6 Comparison between DLVO and XDLVO plots of hectorite suspensions, as a function of ionic strength

Hectorite is a trioctahedral smectite clay mineral with a modest layer charge which is mildly hydrophobic (see Table 8.5 for the pertinent data on this mineral) (van Oss *et al.*, 1990a). A dispersion of 1% (w/v) of hectorite in water of different ionic strengths (with NaCl) showed the following results: 1 M NaCl; metastable (stable for about 1/2 hour, but then instability slowly sets in) and for 0.01M NaCl: stable. Classical DLVO analysis (Figure 8.5) would predict stability up to the highest salt concentrations, which was clearly not realistic. XDLVO analysis, however, predicts instability at $\mu = 1.0$, metastability at $\mu = 0.1$ and rather solid metastability at $\mu = 0.01$ (with a secondary maximum of repulsion of about +80 kT at a distance of about 3 nm (Figure 8.6). Thus, XDLVO analysis is more closely predictive of the experimental outcome of the stability of hectorite as a function of ionic

Parameter	Value
γ^{LW}	39.9 mJ/m^2
$\gamma^{\oplus}$	0.0 mJ/m^2
$\gamma^{\ominus}$	23.7 mJ/m^2
ζ(at 0.015M NaCl)	-38 mV
z (thickness of the slipping plane)	0.5 nm
λ (decay length of water)	1 nm
particle radius	1 μm
$1/\kappa$ (Debye length; i.e., thickness of electrical double layer)	2.53 nm

Table 8.5: *Physico-chemical parameters for the calculation of the free energy of interaction for hectorite.*

strength. Simple DLVO analysis (neglecting all AB interactions) was wrong at all ionic strengths at $\mu \geq 0.1$.

8.9 Influence of Plurivalent Cations on the Flocculation of Negatively Charged Particles: DLVO and XDLVO Analysis

In section 8.4, the linkage between the electrokinetic (ζ-) potential and electron-donicity was discussed. For, e.g., negatively charged particles, a reduction in ζ-potential concomitantly gives rise to a reduction in $\gamma_i^{\ominus}$, thus causing the particles to become hydrophobic, which is the underlying cause of the flocculation of initially negatively charged hydrophilic particles by the admixture of small amounts of Ca^{2+} or of very small amounts of La^{3+} ions, which makes them hydrophobic. Figure 8.7 shows DLVO and XDLVO diagrams of an originally hydrophilic calcite particle suspension, which was turned hydrophobic by the admixture of 3.9 mM $CaCl_2$. The Ca^{2+}-treated calcite particle suspension shown in Figure 8.7 had the properties shown in Table 8.6.

At $\ell = \ell_0$, ΔG_{iwi}^{LW} = -94 kT, ΔG_{iwi}^{AB} = 18,000 kT and ΔG_{iwi}^{EL} = +140 kT (Wu, 1994). A simple DLVO-plot, only taking ΔG_{iwi}^{LW} and ΔG_{iwi}^{EL} into account, shows an extremely short-range van der Waals attraction which, within less than 1 nm distance turns into a long-range electrostatic repulsion, connoting a rather solid metastability. This did not occur; the Ca^{2+}-treated calcite sus-

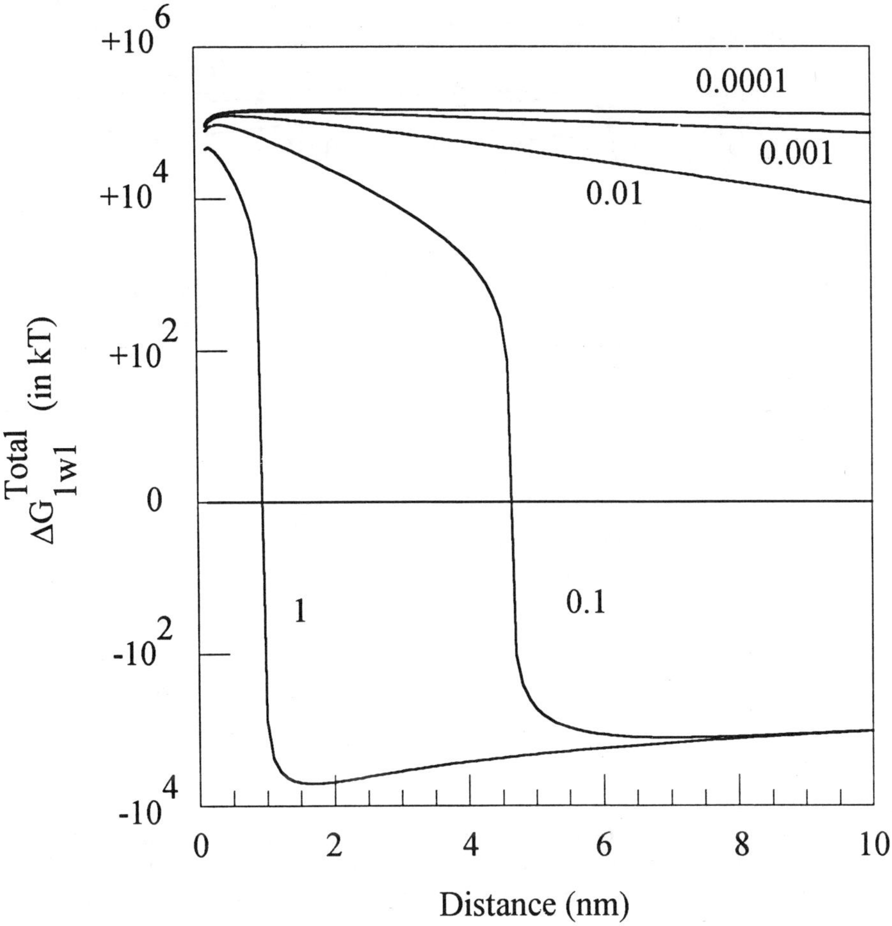

Figure 8.5: *A classical DLVO plot of the free energy of interaction (in kT) between 1 μm spherical particles of hectorite as a function of the interparticle distance (in nm) immersed in water of different NaCl concentrations. These concentrations (in M) are shown on each curve. For all concentrations, at contact, a substantial repulsion between particles would be expected.*

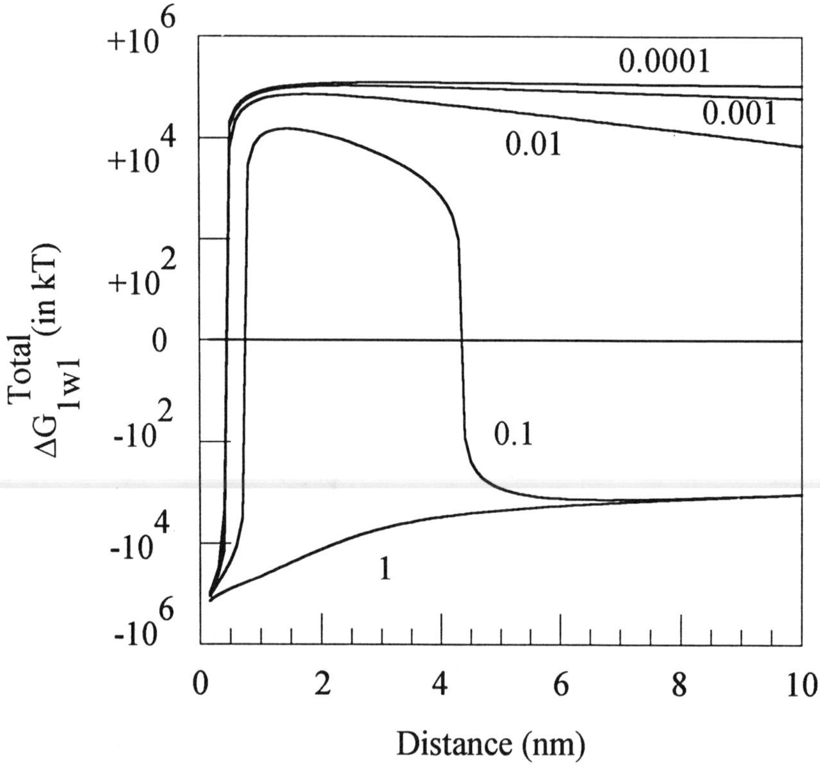

Figure 8.6: *An XDLVO plot of the free energy of interaction (in kT) between 1 μm spherical particles of hectorite as a function of the interparticle distance (in nm) immersed in water of different NaCl concentrations. These concentrations (in M) are shown for each curve. For all concentrations, at contact, there is a substantial attraction between particles. A suspension in 1M NaCl is predicted to flocculate, the 0.1M suspension has an energy barrier at close approach with a secondary minimum beyond 5 nm, while for all other concentrations there is a substantial energy barrier as the particles approach each other.*

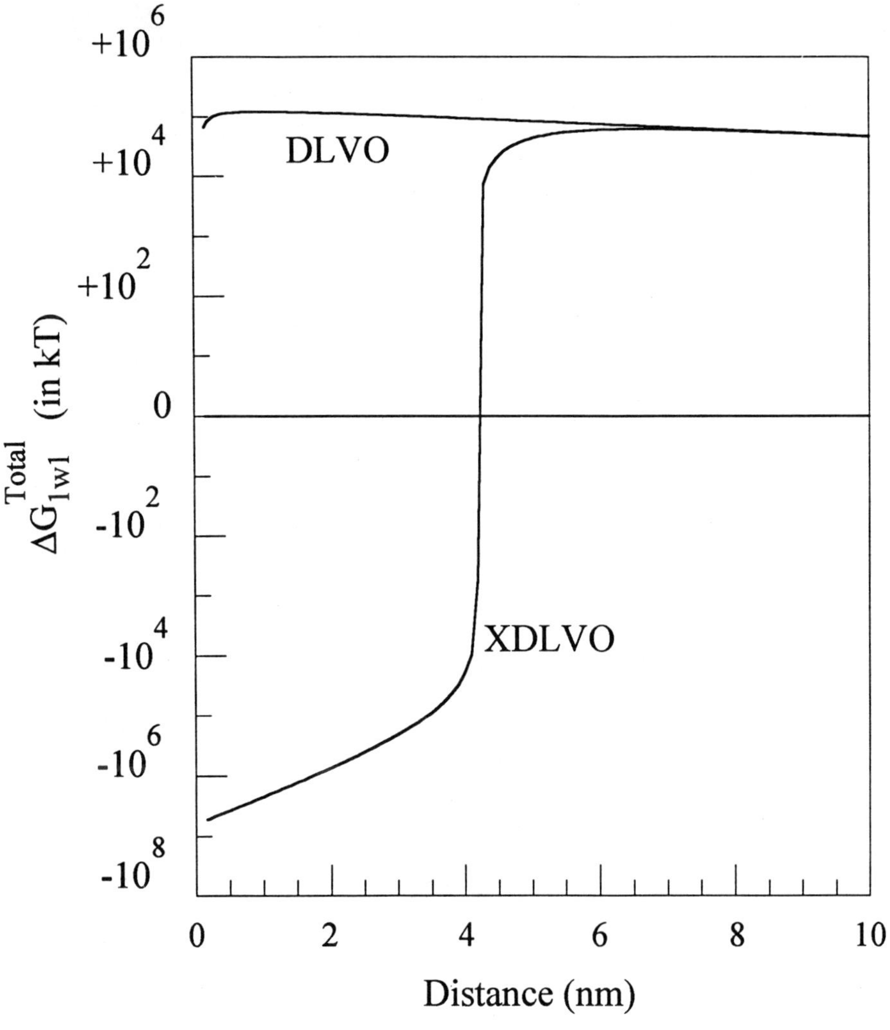

Figure 8.7: *Free energy vs. distance curves obtained with the DLVO and XDLVO theories for calcite particles dispersed in 3.9 mM* $CaCl_2$ *(Wu, 1994).*

Parameter	Value
γ^{LW}	28.0 mJ/m^2
$\gamma^{\oplus}$	0.3 mJ/m^2
$\gamma^{\ominus}$	14.2 mJ/m^2
ζ(at 0.015M NaCl)	-10.6 mV
z (layer thickness at the slipping plane)	0.5 nm
λ (decay length of water)	1 nm
particle radius	1 μm
$1/\kappa$ (Debye length: i.e., thickness of electrical double layer	2.53 nm

Table 8.6: *Physico-chemical parameters for the calculations of the free energy of interaction for calcite hydrophobized by treatment with 3.9 mM $CaCl_2$.*

pension flocculated, due to the massive hydrophobic (AB) attraction shown in the XDLVO diagram (Figure 8.7).

8.10 Solubility

There is no intrinsic difference between the solubility of solutes in a solvent and the stability of particles suspended in a liquid medium. For micron-sized particles and for polymeric molecules with molecular weights of 10^5 or larger, absolute repulsion (i.e., $\Delta G_{iwi} > 0$) is a requirement for stability in aqueous suspensions, or for aqueous solubility. For solutes of low molecular weight, aqueous solubility can still be attained at $\Delta G^{IF}_{iwi} < 0$, provided that the value of ΔG^{IF}_{iwi} in such cases be not much more negative than a few kT; using Eq. 6.19 in the form:

$$\Delta G^{IF}_{iwi} \cdot S_c = kT \cdot \ln s \tag{8.7}$$

then for a solute of molecular weight 100, an aqueous solubility of 10% (w/v), at 20°C, corresponds to ΔG^{IF}_{iwi} = -4.0 kT. (In Eq. 8.7 s is expressed in mole fractions, to find ΔG^{IF}_{iwi} in kT units; 1 mole fraction = 55.56 ML^{-1}); this value assumes the absence of cluster formation among the water molecules. If one wishes to take water clusters into account, these are of the order of approximately 5.1 molecules of water per cluster, at 20°C, so that then 1 mole fraction = 10.9 ML^{-1} (van Oss *et al.*, 2001b). Then a solute of molecular weight 100, when ΔG^{IF}_{iwi} = -5 kT, s = 0.73%; at -6 kT, s = 0.27% and at -7 kT, s = 0.1 % .

Parameter	Value
γ^{LW}	42.0 mJ/m^2
$\gamma^{\oplus}$	0.0 mJ/m^2
$\gamma^{\ominus}$	55.0 mJ/m^2
ζ(at 0.015M NaCl)	0.0 mV
λ (decay length of water)	1.0 nm
particle radius	1 μm

Table 8.7: *Parameters for the calculations of the free energy of interaction for dextran D-150 in water, as shown in Figure 8.8.*

8.10.1 Solubility of electrolytes

From the solubilities of a number of the more common electrolytes (e.g., KCl, NaCl, $(NH_4)_2SO_4$) it can be seen that in many of these cases, ΔG_{1w2}^{Total} has a value of the order of +1.5 kT or more (the subscripts allude to the cation and the anion). For monovalent/monovalent salts, $\Delta G_{1w2}^{LW} \approx$ -0.25 kT. Thus, a total of about +0.2 kT has to be furnished to obtain $\Delta G_{1w2}^{Total} \approx +1.5$ kT, and $\Delta G_{1w2}^{Brownian}$ can furnish only $\approx +1$ kT. Almost 1 kT must therefore be provided by another source, which is ΔG_{1w2}^{AB}. However, naked cations are electron accepters, and anions are electron donors, which *attract* each other and thus would give rise to insolubility. On the other hand, taking hydration orientation into account, cations will be hydrated with the H atoms of the water molecules pointing outward. But small anions will have only one H atom attached to the naked ion, which causes the other H atom (given the H-O-H angle of 104.5°) also to point outward; see also Franks (1984). In this manner, a net AB-repulsion can evolve between small hydrated anions and cations, thus easily providing the missing $\Delta G_{1w2}^{AB} \approx 1$ kT.

Aqueous solubility of electrolytes is less favored in the case of the heavier ions, due to a stronger LW attraction, and especially in the case of plurivalent ions, in particular when both anions and cations are plurivalent.

8.10.2 Solubility of organic compounds

In using the solubility equation (Eq. 8.7) it should be noted that the determination of the contactable surface area (S_c) between two identical molecules requires special attention. The S_c-values of a number of strictly, or mainly, apolar compounds could be determined from the known interfacial tensions with water, and the known solubilities in water, cf. Table 8.8. It is obvi-

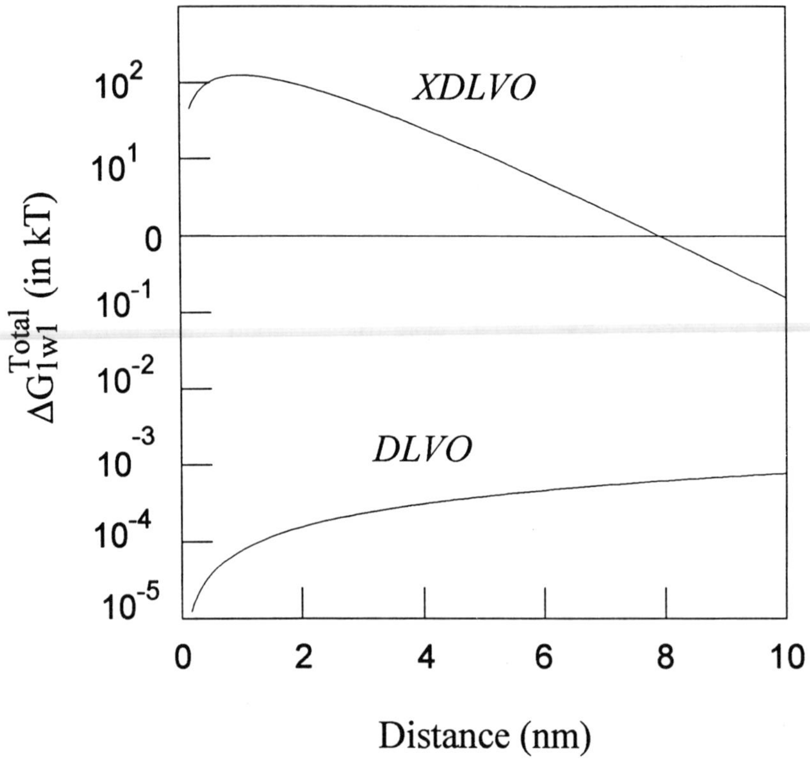

Figure 8.8: *Free energy versus distance curves obtained with the DLVO and XDLVO predictions for D-150 dextran molecules in water; for data concerning these dextran molecules, see Table 8.7.*

ous that the contactable surface area is considerably smaller than the total surface area of these molecules (see, e.g., Tanford (1979)). The contactable surface area between two organic molecules, immersed in water, is the common surface area between solute molecules, at closest approach. The contact is best considered to occur between the surfaces of the Born-Kihara shells of these solute molecules (also called their van der Waals surface, or their solvent cavity surface). For roughly cylindrically-shaped molecules such as alkanes, as a first approximation, S_c appears to be about equal to one sixth of the total Kihara surface area, which agrees rather well with the common surface that may be expected to exist between two parallel adjoining cylinders, analogous to the hexagonal cross-section of honeycomb elements (van Oss and Good,1996). With respect to Table 8.8 it should be noted that in expressing the solubility of organic compounds in water in mol fractions, it is assumed that there are 55.56 moles of water in a liter. However, if one wishes to take the cluster formation of water into account (at approximately 5.1 water molecules per cluster, at room temperature) then instead of 55.56 moles of water per liter, there would be only 10.9 moles of water clusters per liter, so that the solubilities in mole fractions shown in the middle column of Table 8.8 would be 5.1 times higher, which decreases the (absolute) ΔG_{iwi} values by about 1.63 kT, which then also somewhat decreases the resulting S_c-values. See also section 8.1.

The importance should be stressed of always taking ΔG_{iwi}^{AB} into account as an essential part of ΔG_{iwi}^{Total}, in expressing the aqueous solubility of organic compounds, using Eq. 8.7. For example, in the solubility of the alkane, hexadecane ($C_{16}H_{34}$), if one were only to take ΔG_{iwi}^{LW} and ΔG_{iwi}^{EL} into account, one would arrive at total solubility of this compound in water:$\gamma_i^{LW} = 27.5$ mJ/m^2, so that $\Delta G_{iwi}^{LW} = -0.66$ mJ/m^2, which at $S_c = 1.0$ nm^2, amounts to -0.16 kT. The ψ_o-potential of hexadecane, at pH 7.0, in 0.01 M K_3PO_4 is -56 mV (Geertsema-Doornbusch *et al.*, 1993), so that $\Delta G_{iwi}^{EL} = +0.51$ kT. Thus, $\Delta G + \Delta G_{iwi}^{EL} = -0.35$ kT, which allows total solubility, which is erroneous. Also taking $\Delta G_{iwi}^{AB} = -102$ mJ/m^2 into account however, yields $\Delta G_{iwi}^{Total} =$-100.6 mJ/m^2 (at $\ell = \ell_o$), giving rise to $s = 19$ ng/100 ml, indicating virtually total insolubility in water, which is indeed the case.

The solubility of organic polymers in water is also decided by the value of ΔG_{iwi}^{Total}, at $\ell = \ell_o$. A water-soluble polymer, the electrostatically neutral (Edberg *et al.*, 1972) linear polymer of glucose, *dextran*, may be taken as an example. As $\Delta G_{iwi}^{EL} = 0$ (Edberg *et al.*, 1972), one needs only to consider $\Delta G_{iwi}^{LW} = -6.56$ mJ/m^2 and $\Delta G_{iwi}^{AB} = +47.80$ mJ/m^2 (for dextran D-150, van Oss (1994)). In the repulsive mode the S_c for crossed cylinders of the D-150

Compound	γ_{1w} (mJ/m^2)	ΔG_{1w1} (mJ/m^2)	Solubility (exptl.) (mol frac)	ln s (=ΔG_{1w1}) (kT)	S$_c$ (calc.) (nm^2)	ΔG_{1w} (kT)
pentane	49.0	-98.0	9.4 ×10^{-5}	-9.275	**0.382**	-3.768
heptane	50.2	-100.4	9.0 ×10^{-6}	-11.618	**0.468**	-4.750
octane	50.8	-101.6	2.21 ×10^{-6}	-13.023	**0.518**	-5.593
cyclohexane	50.0	-100.4	11.12×10^{-6}	-11.407	**0.450**	-5.435
benzene	35.0	-70.0	4.038×10^{-4}	-7.815	**0.451**	-7.440
naphthalene	39.2	-78.4	4.14 ×10^{-6}	-12.395	**0.639**	-10.502
CS_2	48.4	-96.8	2.63 ×10^{-4}	-8.244	**0.344**	-4.828
CCl_4	45.0	-90.0	9.29 ×10^{-5}	-9.284	**0.417**	-5.656
chloroform[a]	38.8	-77.6	12.06×10^{-4}	-6.720	**0.350**	-5.291

Table 8.8: *Surface properties, solubilities in water, free energies of interaction and contactable surface areas of strictly apolar and mainly apolar solutes (van Oss and Good, 1996).* [a] *From van Oss et al. (2001a).*

molecule, amounts to about 1.88 nm^2, which yields a value for ΔG_{iwi}^{Total}, at $\ell = \ell_o$ of +19.2 kT, yielding a solubility, s, of 2.2 × 10^8 mol fractions, indicating total solubility in water, as is indeed the case. However, if one chooses to neglect the Lewis acid-base interactions (which are crucial in all interactions in water) then one would only consider ΔG_{iwi}^{LW}. In the attractive mode, the D-150 dextran molecule, which is 1.37 nm wide and 100 nm long (van Oss 1994) could theoretically have a contactable surface of 68.5 nm^2 (Edberg *et al.*, 1972). This would lead to ΔG_{iwi}^{LW} = -111 kT (in the attractive mode). This value would correspond to complete insolubility (cf. Eq. 8.7) which is obviously wrong, given the total solubility of dextran in water.

Thus, for the aqueous solubility of organic monomeric or polymeric compounds, ΔG_{iwi}^{AB} should always be taken into account, irrespective of whether one has to do with practically insoluble compounds (such as hexadecane), or with exceedingly soluble molecules (such as dextran).

8.10.3 Solubility of surfactants and other amphipathic compounds

Whilst the connection between the aqueous solubility of alkanes and similar apolar compounds, and the interfacial tension of such compounds with water, γ_{iw} (where $-2\gamma_{iw} = \Delta G_{iwi}^{IF}$), allows the direct utilization of Eq. 8.7, the use of that equation is not directly allowable where surfactants or other amphipathic compounds are concerned. In other words, where Eq. 8.7 works very well for alkanes such as heptane, octane, etc. (cf. Table 8.8), for heptanol and octanol, the solubilities calculated via Eq. 8.7 from their γ_{iw} and S_c-values, are hundreds of times higher than the experimental solubilities. This is because the experimentally measured γ_{iw} values, e.g., by means of the drop-shape of the organic compound in water (cf. Section 6.10.2) measures a skewed γ_{iw}, caused by the fact that, e.g., in the case of alkanols, the OH-group is oriented toward the water interface, so that one only measures apparent γ_{iw}-values, which are about a decimal order of magnitude too small. Nonetheless, there is a way of using Eq. 8.7, which is by measuring the γ_{iw}-values of the apolar moiety and of the polar moiety of the γ_{iw} *separately*, calculating the *separate* ΔG^{IF}_{iwi}-values from these, expressing them in kT units and then *adding them together*, resulting in a final $\overline{\Delta G^{IF}_{iwi}}$-value, in kT units (van Oss and Good, 1996).

The same can be done with ionic and nonionic surfactants, allowing the precise prediction of the critical micelle concentration (cmc), which for surfactants is equal to their solubility (Adamson, 1990; van Oss 1994).

8.11 Adhesion and Adsorption

For aqueous systems, the free energy of interaction between particle or molecule, 1, and solid substratum, 2, immersed in water, w, is described by (cf. Eq. 6.50):

$$\begin{aligned}\Delta G_{1w2} &= \sqrt{\gamma_1^{LW}\gamma_w^{LW}} + \sqrt{\gamma_2^{LW}\gamma_w^{LW}} - \sqrt{\gamma_1^{LW}\gamma_2^{LW}} - \gamma_w^{LW} \\ &+ 2\sqrt{\gamma_w^{\oplus}}\left(\sqrt{\gamma_1^{\ominus}} + \sqrt{\gamma_2^{\ominus}} - \sqrt{\gamma_w^{\ominus}}\right) \\ &+ 2\sqrt{\gamma_w^{\ominus}}\left(\sqrt{\gamma_1^{\oplus}} + \sqrt{\gamma_2^{\oplus}} - \sqrt{\gamma_w^{\oplus}}\right) \\ &- 2\sqrt{\gamma_1^{\oplus}\gamma_2^{\ominus}} - 2\sqrt{\gamma_1^{\ominus}\gamma_2^{\oplus}}\end{aligned} \qquad (8.8)$$

Eq. 8.9 applies equally to adhesion and to adsorption, and also to repulsive interactions.

When, in addition to ΔG^{IF}_{1w2}, the electrostatic interaction energy must be taken into account, the ΔG^{EL} versus distance equation given in Table 5.1 should be used. The value of ψ_o is then obtained from the values $\psi_{o(1)}$ and $\psi_{o(2)}$ by means of the geometric mean combining rule:

$$\psi_{o(12)} = \sqrt{\psi_{o(1)}\psi_{o(2)}} \qquad (8.9)$$

8.11.1 Macroscopic scale adhesion and adsorption

This is the type of adhesion occurring with, e.g., hydrophilic particles or macromolecules adhering onto a hydrophilic surface by a strictly (or largely) hydrophobic attraction. Examples are: adhesion of bacteria (*Pseudomonas aeruginosa*) onto hydrophobic dolomite particles where the least hydrophilic bacteria (i.e., the ones harvested in the stationary growth phase) adhere most strongly to the dolomite particles (Grasso *et al.*, 1996); adsorption of proteins, onto hydrophobic surfaces (MacRitchie, 1972; van Oss 1995); and, finally, adhesion and adsorption onto the water-air interface (see flotation in the following section).

All types of straightforward macroscopic-scale adhesion and adsorption can be completely described by the equations given in Table 8.4, and by simple XDLVO diagrams.

8.11.2 Adhesion and adsorption onto the water-air interface - flotation

The air-side of the air-water interface, with $\gamma^{LW} = 0$, $\gamma^{\oplus} = 0$ and $\gamma^{\ominus} = 0$, is perfectly hydrophobic, i.e., at 20°C, the acid-base interaction energy between two air surfaces (e.g., two air bubbles), in pure water, ΔG^{AB}_{iwi} = -102 mJ/m^2. However, typical mineral or biological materials of $\gamma^{LW} \approx 40$ mJ/m^2, undergo an LW repulsion by an air interface, in water, of $\Delta G^{LW}_{iwi} \approx +5.5$ mJ/m^2, resulting in a total hydrophobic attraction, $\Delta G^{IF}_{iwi} \approx$ -96.5 mJ/m^2. Thus, the adsorption of biopolymers, such as proteins, onto the water-air interface plays a considerable role in the air-drying of such materials. Drying a protein solution for instance causes the protein molecules to orient their (normally internal) hydrophobic moieties toward the air interface, so that the final dried material is *hydrophobic*, even when the native protein is hydrophilic (van Oss 1994). For instance, (human) serum albumin and serum immunoglobulin G are hydrophobic in the dried state and hydrophilic in the hydrated state, so that dried powders of these proteins when put into water, do not at first dissolve, but after about a minute or so, slowly take up water, rehydrate and then suddenly dissolve completely. Less globular (more elongated) proteins such as fibrinogen, do not show this behavior.

All hydrophobic and many hydrophilic materials are attracted to the air-side of the air-water interface. In order to acquire a very large water-air interface, the number of air bubbles can be greatly enhanced by the admixture of surfactants. However, the advantage of a greater water-air interface, entails the disadvantage of a decrease in hydrophobicity of that interface due to the added surfactants. In any event, the use of surfactants in large scale flotation is widespread, e.g., in the separation of limestone (in the foam phase) from clay particles (which sediment) (Marks, 1971), in the

Material	γ^{LW} (mJ/m^2)	$\gamma^{\oplus}$ (mJ/m^2)	$\gamma^{\ominus}$ (mJ/m^2)	ψ_o (mV)	ΔG_{1w2} [a] (mJ/m^2)
glass	33.7	1.3	62.2	-59.3	30.33
HSA (2 layers of water)	26.8	6.3	50.6	-31.8	

[a] ΔG^{LW}_{1w2} = -1.15; ΔG^{AB}_{1w2} = +30.55; ΔG^{EL}_{1w2} = +0.93 mJ/m^2.

Table 8.9: *Surface properties of clean glass and of hydrated human serum albumin (HSA) with two layers of hydration (van Oss 1994; van Oss et al., 1995b), and their close-range macroscopic-scale interaction energy.*

separation of calcite, quartz and limestone (in the foam phase) from fluorspar and metal sulfides (van Thoor, 1971), and in the separation of graphite (in the foam phase) from clay particles (Smith, 1971). In the latter case the liquid medium consists of a mixture of water and kerosene, with pine oil (a mixture of terpene alcohols) as surfactant. This surfactant is also used in the flotation of lead and zinc ores (Merck Index, 1989).

8.11.3 Macroscopic and microscopic-scale adsorption phenomena combined

On a macroscopic level, hydrophilic biopolymers (e.g., serum albumin), dissolved in water, undergo a strong hydrophilic (AB) and a weak electrostatic (EL) repulsion at neutral pH, vis-a-vis a hydrophilic surface such as clean glass, which should prevent such biopolymers to make contact with a glass surface and to become adsorbed to it. Nonetheless, adsorption of hydrophilic biopolymers onto hydrophilic mineral surfaces (such as glass) occurs quite commonly. The mechanism of this is twofold: adsorption takes place as the final outcome of two opposing interactions: macroscopic-scale repulsion versus microscopic-scale attraction. The macroscopic repulsion between e.g., human serum albumin (HSA) and a clean glass surface is easily determined; see Table 8.9. Mainly due to the strong hydrophilic (AB) repulsion, the total repulsive energy between HSA and clean glass is about +30 mJ/m^2, which at first sight would seem to preclude any approach between the two materials to within several (3 or 4) nm. However, imbedded in the glass are numerous plurivalent cations (most often Ca^{2+}), which often have an excess positive charge at the glass surface; such a positive charge also acts as an electron-acceptor. Then a moiety on the protein's surface, with a small radius of curvature, R, can: 1) penetrate the general repulsive field more deeply than straight peptide moieties (cf. Table 8.4 for the influence of R), and 2) being negatively charged and a strong electron-donor, will approach, and attach itself to a discrete Ca^{2+}-containing site on the glass surface. The same mechanism applies to protein adsorption onto ground glass particles and to clay and other mineral (metal oxide) particles which all tend to have plurivalent metal ions at the surface, especially at sharp edges arising from cleavage or fracture.

In this manner, hydrophilic proteins can adsorb onto equally hydrophilic metal oxide surfaces, even though the total adsorption avidity is somewhat lower than with adsorption onto hydrophobic surfaces (MacRitchie, 1972). A strong indication that the above scenario, which invokes the adsorption of proteins to discrete sites comprising plurivalent cations, is correct, is the experimental finding that proteins, adsorbed onto metal oxide surfaces, can be desorbed by the admixture of complexing agents, such as Na_2EDTA or hexametaphosphate (van Oss *et al.*, 1995a,b). It may be assumed that the vast majority of cases of adsorption of organic biopolymers onto clay and other mineral particles, in nature, follows this mechanism.

8.11.4 Adsorption and adhesion kinetics

Based on von Smoluchowski's (1917) equation for the flocculation kinetics of small particles, the kinetic adsorption constant (k_a) can be expressed as:

$$k_a = 4\pi \ell_o D f \frac{N}{1,000} \tag{8.10}$$

where ℓ_o is the minimum equilibrium distance ($\ell_o = 0.157$ nm; cf. Section 5.3) between two molecules or particles that are not interacting covalently, D is the diffusion constant of the adsorbing solute, A is Avogadro's constant ($N = 6.02 \times 10^{23}$) and f is von Smoluchowski's net probability factor, incorporating both the improbability of an encounter between, e.g., protein and substratum (originating in the macroscopic repulsion), and the probability of such an encounter (driven by the microscopic attraction):

$$f = \int_\phi \left\{ \exp\left[\frac{1}{\ell - \ell_o} \left(\int_{\ell=\ell_o}^{\ell=\infty} \frac{-\Delta}{kT} d\ell \right) \right] \right\} d\phi \tag{8.11}$$

so that f embodies all interaction energies (repulsive and attractive), at all distances, ℓ, from $\ell = \ell_o$, to $\ell = \infty$, and at all orientations (ϕ) of the adsorbing solute (van Oss 1997). To determine the f-factor (Eq. 8.11), one needs to use the XDLVO diagrams of both the macroscopic repulsion situation(s) for those orientations, ϕ_1, where repulsion prevails, and the net microscopic attraction case(s), for the (usually somewhat rarer) orientations, ϕ_2, leading to attraction. Based on the data measured for the kinetics of the adsorption of human serum albumin (HSA) onto polydispersed silica particles (Docoslis *et al.*, 2000b) the macroscopic (repulsive) part of the value $\frac{\Delta G^{mac}}{kT}\frac{d\ell}{\ell} = \chi^{mac}$ (integrated and averaged over all distances from ℓ_o to ℓ = 10 nm, beyond which no significant measurable interaction occurs): χ^{mac} = +14.66 kT. The microscopic (attractive) part of $\frac{\Delta G^{mic}}{kT}\frac{d\ell}{\ell} = \chi^{mic}$ (also integrated and averaged from ℓ_0 to ℓ = 10 nm: χ^{mic} = -2.96 kT. Then, from these data and from χ^{Tot}:

$$\chi^{tot} = a\chi^{mac} + (1-a)\chi^{mic} \tag{8.12}$$

where

$$\chi^{Tot} = \ln f \tag{8.13}$$

is the unfavorable to favorable orientation ratio $r(\phi) = a/(1-a)$ could be determined to be 1.46 ± 0.4 (SD), and (Eq. 8.10) could be determined from the measured adsorption rate constant, $k_a = 9.74 \times 10^4$ L/Ms, inserted into Eq. 8.10, from which it follows that $f = 1.37 \times 10^{-3}$ (Docoslis *et al.*, 2001). This all means that the average repulsive energy is about a factor:

$$R = 1.46 \times \frac{\chi^{mac}}{\chi^{mic}} = 7.23 \tag{8.14}$$

7.23 times more energetic than the average energy of attraction, but the kinetic adsorption rate constant is $1/f$ = 730 times slower than it would have been in the absence of marcoscopic-scale repulsion.Thus, the influence of the macroscopic-scale background repulsion on the kinetic adsorption constant is enormous, in all cases of adsorption of a hydrophilic polymer to a hydrophilic substratum, in water. However, as the kinetic dissociation constant, k_d, is expressed as:

$$k_d = (4\pi\ell_o D f \frac{N}{1,000})/K_{eq} \tag{8.15}$$

so that:

$$K_{eq} = k_a/k_d \tag{8.16}$$

(cf. Eq. 8.10) the equilibrium association constant, K_{eq}, is not affected by the factor, f; see below. When for a given adsorption system of, e.g., a protein, adsorbing onto a clay particle, the kinetic adsorption rate constant, k_a has been ascertained, von Smoluchowski's f-factor can also be determined via Eq. 8.10, if the protein's diffusion constant, D, is known.

8.11.5 Adsorption and adhesion equilibrium

The equilibrium adsorption constant, K_a, is entirely independent of the factor, f (cf. Eq. 8.16), so that the value of K_{eq} is not influenced by the background repulsion, but is only a function of the free energy of attraction at the minimum equilibrium distance, $\ell = \ell_o$, because K_{eq} can also be expressed as:

$$K_{eq} = \exp \frac{-\Delta G_{(\text{at } \ell=\ell_o)}}{kT} \tag{8.17}$$

where K_{eq} can be found as the slope of the Langmuir adsorption isotherm (plotting the amount adsorbed versus the solute concentration), at zero solute concentration.

To avoid the pervasive influence of steric hindrance of already adsorbed molecules, which prevents the attraction between hampered sites, and other solute molecules that are still in solution, both kinetic adsorption and desorption constants, k_a and k_d, must be measured after exceedingly short times (typically within a few tenths of a second) and at the lowest solute concentrations compatible with adequate mass transport.

8.12 Net Repulsive Interactions

8.12.1 Reversal of adsorption and adhesion

Adsorption or adhesion, when based on non-covalent attractive forces, can usually be reversed by: applying conditions of extremes of pH; displacement with amphipathic agents (e.g., surfactants, acetonitrile, etc.); or for metal ion-mediated microscopic level adsorption, displacement by competing agents, such as closely resembling proteins (in the case of proteins) or ligands, such as complexing agents.

EXTREMES OF pH are typically achieved with: 10 mM HCl (pH $\approx$ 2); 50 mM NaOH (pH $\approx$ 10) (Mayers and van Oss, 1998) or 70% formic acid.

DISPLACEMENT WITH AMPHIPATHIC AGENTS is especially effective for removing adsorbed macromolecules from solid substrate, but it often is difficult to remove surfactants from the desorbed macromolecules. Acetonitrile, which is widely used for the desorption step in reversed-phase liquid chromatography, works in the same way as surfactants, but higher concentrations are needed. However, concentrations higher than 30% are no longer effective: through dehydration effects such high concentrations of acetonitrile re-solidify the adsorption of biopolymers onto hydrophobic carriers.

COMPETING AGENTS are especially effective in displacing proteins that have been microscopically adsorbed onto metal oxides (van Oss *et al.*, 1995b) or metal ions imbedded in glass surfaces (van Oss *et al.*, 1995a), and equally effective are very similar proteins, such as for instance a given radioactive protein, which is very effectively replaced by the same, non-radioactive protein; see, e.g., Ball *et al.* (1996).

8.12.2 Temperature effects

In adhesion and adsorption in aqueous media, temperature effects can play an appreciable role. This is because the $\gamma_{w}^{\oplus}$ and $\gamma_{w}^{\ominus}$ values for water, which are fixed at 25.5 and 25.5 mJ/m^2/ at 20°C (cf. Chapter 6), change rather drastically with temperature. Upon increasing the temperature, water becomes more Lewis acidic and less Lewis basic. For instance, at 38°C, the γ-values for water are: $\gamma_{w}^{LW} = 21.0$, $\gamma_{w}^{\oplus} = 32.4$ and $\gamma_{w}^{\ominus} = 18.5$ mJ/m^2 (van Oss 1994); the $\gamma_{w}^{\oplus}/\gamma_{w}^{\ominus}$ ratio for water thus changes from 1.0 at 20°C, to 1.75 at 38°C. As shown in Table 8.10, the influence of an 18°C increase in temperature has only a slight influence on the adsorption of a partly polar substance, amylopectin (from corn starch) onto a completely hydrophobic substratum (Teflon), but there is an appreciable change in adsorption energy of amylopectin onto a partly polar substratum, cellulose acetate, and there is a change from adsorption to desorption upon heating from 20° to 38°C in the case of the interaction between amylopectin and the mildly hydrophilic substratum, dolomite. These surface thermodynamic predictions agree well with a phenomenon that almost everybody has noticed at one time or another: it is much easier to clean cooking pots and pans and soiled dishes in hot water than in cold water, even in the absence of soap or other surfactant.

It is clear that the major changes in ΔG_{1w2} as a function of temperature, are due to changes in ΔG_{1w2}^{AB}. With adsorption of amylopectin onto cellulose acetate, ΔG_{1w2}^{AB} turns repulsive at 38°, but not quite enough to surmount the LW attraction. With dolomite as a substratum, the AB repulsion at 38°C is largely sufficient to turn the attraction at 20°C into a net repulsion at 38°C; cf. Table 8.10.

Adsorption of amylopectin onto	Temp. °C	ΔG^{LW}_{1w2} mJ/m^2	ΔG^{AB}_{1w2} = mJ/m^2	ΔG^{IF}_{1w2} mJ/m^2
Teflon	20	+1.5	-44.5	-43.0
	38	+1.5	-35.5	-34.0
cellulose acetate	20	-6.8	-11.1	-17.9
	38	-7.6	+5.1	-2.5
dolomite	20	-4.9	-2.0	-6.9
	38	-4.6	**+14.0**	**+9.4**

[a] amylopectin:$\gamma_1^{LW} = 40.1; \gamma_1^{\oplus} = 0.4; \gamma_1^{\ominus} = 18.5$ mJ/m^2 (van Oss 1994)
[b] Teflon:$\gamma_2^{LW} = 17.9; \gamma_2^{\oplus} = 0; \gamma_2^{\ominus} = 0$ mJ/m^2 (van Oss 1994)
[c] cellulose acetate:$\gamma_2^{LW} = 44.9; \gamma_2^{\oplus} = 0.8; \gamma_2^{\ominus} = 23.4$ mJ/m^2 (van Oss 1994)
[d] dolomite:$\gamma_2^{LW} = 37.6; \gamma_2^{\oplus} = 0.2;\ \gamma_2^{\ominus} = 30.5$ mJ/m^2 (Giese *et al.*, 1996)

Table 8.10: *Interfacial free energies (ΔG^{IF}_{1w2}) of adsorption (or desorption) of amylopectin[a] onto Teflon[b], cellulose acetate[c] and dolomite[d] surfaces, in mJ/m^2, as a function of temperature using Eq. 8.9.*

8.13 Nature of Clay-water Interactions

The free energy of hydration of the exterior surface of clay and other mineral particles can be determined by means of the Dupré equation:

$$\Delta G_{iw} = \gamma_{iw} - \gamma_i - \gamma_w \tag{8.18}$$

The free energy of hydration of the outer surface of clay particles varies from about ΔG_{iw} = -85 mJ/m^2/ for (the very hydrophobic) talc surface, to ΔG_{iw} = -126.5 mJ/m^2 for the mildly hydrophilic smectite, SWy-1 and ΔG_{iw} = -143 mJ/m^2 for the very hydrophilic smectite, SAz-1 (Giese *et al.*, 1996); see also the ΔG_{iw}-values for various other materials in Table 8.3. For hydrophobic materials the hydration energy, ΔG_{iw} (cf. Table 8.11) is close to 100% due

	γ^{LW}	$\gamma^{\oplus}$	$\gamma^{\ominus}$	γ^{LW}	γ^{AB}	ΔG^{LW}_{1w}	ΔG^{AB}_{1w}	ΔG^{Tot}_{1w}	% LW	% AB
Teflon	17.9	0	0	51	0.19	-38.9	0	-39.5	99.5	0.5
PEO	43	0	64	3.57	-29.8	-61.2	-80.8	-142.0	43.1	56.9

Table 8.11: *Surface properties of Teflon and polyethylene oxide (PEO) and their free energies of hydration.*

to van der Waals forces (cf., e.g., Teflon), while for hydrophilic materials LW-forces account for about 43% (e.g., PEO, Table 8.11).

8.14 Structure of Bound Outer Layer Water

All condensed-phase matter, immersed in water (or to a lesser degree, exposed to water vapor), binds water. Even the most hydrophobic materials (see, e.g., octane, Table 8.12) binds water with a definitely non-negligible energy (proving that the term "hydrophobic" is a misnomer, when taken to mean a repulsion between the material and water). Completely "hydrophobic" materials bind water only with Lifshitz-van der Waals forces (Table 8.12). The most hydrophilic materials (for example, the mica margarite, Table 8.12) bind water with slightly more Lewis acid-base than Lifshitz-van der Waals forces. On all the more or less polar materials shown in Table 8.12 (from mica to pyrophyllite) the AB-bound water is organized (in its first layer) so that the H atoms of water bind to the polar part of the surface, whilst the O atoms stick out into the direction of the bulk water away from the solid surface. Thus, AB-type bound water is oriented and less dense than bulk water. LW-type bound water is unorganized but slightly denser than bulk water. When, e.g., clay particles are insufficiently dried before measuring their surface properties, they manifest a non-negligible value for their $\gamma_i^{\oplus}$, in addition to a more or less sizeable $\gamma_i^{\ominus}$ value. When polar surfaces are well-dried, they only show a significant $\gamma_i^{\ominus}$, i.e., they then are "monopolar" in $\gamma^{\ominus}$ (van Oss *et al.*, 1997). Particles whose water of hydration is denser than the bulk water, attract each other when immersed in water, when the water of hydration is less dense than the bulk water, they repel each other (Derjaguin, 1989).

8.15 Swelling of Clays

8.15.1 Nature of the swelling mechanism

Swelling clay minerals (e.g., smectites) consist of multiple layers of electron-donating, negatively charged layers, with an interlayer of charge compensating cations (e.g., Na^+). Immersed in water and in a number of other polar liquids (e.g., formamide, ethylene glycol, dimethylsulfoxide) smectite clay particles swell, i.e., the distance of about 0.32 nm between two platelets can increase to, e.g., 1.2 nm, or in some cases (with Na^+ and Li^+) the individual

Material	ΔG^{LW}_{iw}	$+ \Delta G^{AB}_{iw}$	$= \Delta G^{IF}_{iw}$	% LW	% AB
Mica (margarite, Chester, MA)	-60.5	-81.8	-142.4	43.0	57.0
Smectite: SWy-1	-59.6	-66.9	-126.5	47.0	53.0
Smectite: hectorite	-59.0	-49.2	-108.2	54.5	45.5
Talc (Fisher Scientific, dehydrated at 400°C)	-54.6	-29.7	-84.3	65.7	35.3
Pyrophyllite	-54.4	-35.5	-89.9	60.5	39.5
Octane	-43.4	-0.2	-43.6	99.5	0.5

Table 8.12: *Free energy of hydration (ΔG^{IF}_{1w}) of a number of clay minerals, ranging from very hydrophilic (mica) to quite hydrophobic (talc and pyrophyllite), with, for comparison, also a 100% hydrophobic condensed-phase material, octane (in mJ/m²); see also Table 8.11.*

clay layers can separate to very large distances or completely disperse in water. Taking a typical smectite (Giese *et al.*, 1996), with $\gamma_i^{LW} = 43.6$, $\gamma_i^{\oplus} = 1.8$ and $\gamma_i^{\ominus} = 36.8$ mJ/m², and a surface potential, ψ_o= -40 mV, with the interstitial surface neutralized with 0.66 Na^+ ions per 0.45 nm, one has 1.47 × 10^6 Na^+ ions per μm^2. Then, per μm^2, one has $\Delta G^{EL}_{Na^+} = +10.5 \times 10^6$ kT/μm^2 for the Na^+-layer (assuming the hydrated radius of an Na^+-ion to be 0.1 nm, and at a thickness of the diffuse double layer, $1/\kappa = 0.1$ nm). From ψ_o= -40 mV (ζ= -31.6 mV) for a smectite layer, one arrives at $\Delta G^{EL}_{smectite} = -4.03 \times 10^6$ kT/μm^2. Then each smectite layer of 1 μm^2, slightly over-compensated with half the available Na^+-ions, will have an electrical energy $\Delta G^{EL}_{iwi} = -1.85 \times 10^6$ kT/μm^2, and $\Delta G^{AB}_{iwi} - 2.89\times 10^6$ kT/μm^2, upon estimation of $\gamma_i^{\oplus} = 0$ and $\gamma_i^{\ominus} \approx 20$ mJ/m², for a smectite layer neutralized with Na^+ ions (cf. Wu (1994)). Thus, between two smectite layers, 0.32 nm apart and 1 μm^2 surface area, neutralized with 1.47 × 10^6 Na^+ ions, there is a strong attraction of $\Delta G^{Total}_{iwi} = -3.5 \times 10^6$ kT; see Table 8.11.

Nonetheless, this attraction can be overcome, at least in part, by swelling, and under some conditions completely, resulting in layer separation. It should be realized that a neutralizing plane of cations, e.g., 1.47 × 10^6 Na^+ ions per μm^2, corresponds to a Na^+ concentration amounting to a supersaturated 7.63 M, causing an osmotic pressure $\Pi = 190$ Kg cm^{-2}. At the edges of the smectite layers, where all the hydrated Na^+ ions are enclosed in the 0.32 nm wide

	$\ell = \ell_o = 0.157$ nm $\mu = 7.63$ $1/\kappa = 0.1$ nm	$\ell = 0.32$ nm $\mu = 7.63$ $1/\kappa = 0.1$ nm	$\ell = 1.20$ nm $\mu = 1.0$ $1/\kappa = 0.3$ nm	$\ell = 1.20$ nm $\mu = 0.001$ $1/\kappa = 10$ nm
ΔG_{iwi}^{EL}	1.22×10^6	1.59×10^5	22.3×10^3	1.08×10^6
ΔG_{iwi}^{LW}	-1.85×10^6	-4.45×10^5	-5.86×10^4	-5.86×10^4
ΔG_{iwi}^{AB} [a] ($\gamma_i^\ominus = 10$ mJ/m^2)	-2.89×10^6	-2.45×10^6	-1.02×10^6	-1.02×10^6
ΔG_{iwi}^{Total}	-3.53×10^6 kT	-2.74×10^6 kT	-1.06×10^6 kT	$+1.4 \times 10^3$ kT

[a] it is assumed that $\gamma_i^\ominus$ remains unchanged upon dilution

Table 8.13: *Interaction energies (ΔG in kT) between smectite layers of 1 μm^2, compensated with Na^+ ions, at different ionic strengths and at ℓ = 0.157, 0.32 and 1.20 nm; $R_{hydrated}$ for Na^+ = 0.1 nm.*

gallery there is thus a tremendous attraction for pure solvent (especially water), but also for other strongly electron-donating solvents, such as glycerol, ethylene glycol, formamide, dimethyl sulfoxide, which have a strong attraction for electron-acceptors, such as concentrated planes of hydrated Na^+ ions. From Table 8.11 it can be seen that when upon swelling, due to interaction with water, the interlayer distance increases from 0.32 nm to 1.20 nm, a net attraction of ΔG_{iwi}^{Total} = -1.076 kT per μm^2 persists, under conditions of high ionic strength (the counterions become somewhat diluted, but a fairly high ionic strength still prevails, so that the thickness $1/\kappa$, of the electrical double layer is still rather small). However, upon lengthy exposure of the smectite to distilled water, many of the interstitial ions can ultimately diffuse out, which causes a significant increase in the thickness of the double layer (here to $1/\kappa$ = 10 nm; see Table 8.11), giving rise to a much decreased decay in ΔG_{iwi}^{EL} as a function of distance, thus allowing the platelets to separate altogether, once a swelling distance of 1.20 nm is achieved; see Table 8.11, right hand column. Nonetheless this is only likely to happen with the smallest monovalent cations (i.e., Li^+, Na^+) and not with the less hydrated monovalent cations (K^+, Rb^+, Cs^+), which have a higher charge density in the hydrated form, and certainly not with plurivalent cations.

8.15.2 Prerequisite properties of swelling clays

As can be seen from Table 8.13, a major prerequisite for the surface of smectite or vermiculite clay particles to allow them to swell, is a negative value for ΔG_{iwi}^{AB} (saturated with Na^+ ions), which is somewhat less negative than the value for ΔG_{iwi}^{EL} under conditions of low ionic strength. Such a relatively moderate negative value for ΔG_{iwi}^{AB} (when saturated with Na^+ ions) is only attainable when the non-neutralized value for $\gamma_i^{\ominus}$ is close to, or larger than about 28 mJ/m^2 (cf. Section 8.7). In other words, given that the inner surface of a smectite platelet has surface properties comparable to those of its outer surface, smectites, to fall into the category of swelling clays, have to be hydrophilic; cf. (Giese *et al.*, 1996), and Section 8.7. Hectorite may appear to be borderline hydrophobic but this is compensated by its ζ-potential of about -44 mV, which actually makes hectorite hydrophilic under conditions of low ionic strength.

Many swelling clays readily form smooth films either by slow drying of an aqueous suspension on a glass slide or on a plastic film. This makes it possible to determine the clay surface tension properties with direct contact angle measurement (cf. Section 6.5.2), but at the same time the tendency to swell prohibits the use of wicking for measuring the surface properties of swelling clays (Section 6.7.1).

The swelling property of smectites is also important for constructing confining barriers aimed at preventing toxic or radioactive waste from leaking beyond their intended bounds. For the containment of toxic products this should, and does, work rather well. See, however, the following section.

8.15.3 Influence of steam on swelling clays

For radioactive waste, which still is capable of generating heat, a problem exists: it has been found that exposure of swelling clays to steam causes the smectite to change from being hydrophilic, to become hydrophobic (Couture, 1985; Bish *et al.*, 1998). In the study of Bish *et al.* (1998) two Na-smectites were examined. Their $\gamma_i^{\ominus}$ values decreased from about 33-34 mJ/m^2, to 21-24 mJ/m^2 after steam treatment, and concomitantly, their ζ-potentials decrease by about 20%. As an example, the surface properties of the smectite SWy-1, at $\gamma_i^{LW} = 41.1$, $\gamma_i^{\oplus} = 1.9$ and $\gamma_i^{\ominus} = 33.2$ mJ/m^2, changes after steam treatment with steam, to: $\gamma_i^{LW} = 42.1$, $\gamma_i^{\oplus} = 1.6$ and $\gamma_i^{\ominus} = 24.2$ mJ/m^2. This means that the initially hydrophilic smectite ($\Delta G_{iwi}^{IF} = +4.4$ mJ/m^2), upon steam treatment becomes hydrophobic ($\Delta G_{iwi}^{IF} = -8.6$ mJ/m^2) (Bish *et al.*, 1999). One striking difference, visible with the naked eye, is that the original SWy-1 sample forms a stable suspension in water, but the same clay after treatment with steam, flocculates upon immersion in water (Bish *et al.*, 1999). Thus, the action of steam on an initially hydrophilic, gel forming, low-permeability material turns it into a hydrophobic, flocculated, porous, granulated material, which has become more permeable, thereby posing problems of containment.

The explanation of the action of steam on smectites appears to lie in the migration of originally tetrahedral Al^{3+} to the clay platelets' external surfaces (Ashton *et al.*, 1984) through displacement with H_2O in the guise of steam, which is strongly Lewis acidic. The attachment of Al^{3+} ions to the clay particles' surface explains the reduction in the particles' (negative) ζ-potential after steam treatment. The interaction between negatively charged surfaces and plurivalent cations is known to have a strong hydrophobizing effect by decreasing a surface's $\gamma_i^{\ominus}$-value, at the same time as its ζ-potential (Wu *et al.*, 1994a); see also Section 8.5.

8.15.4 Hydrophobicity of talc and pyrophyllite

In spite of the fact that both hydrophilic and hydrophobic clay and other minerals particles' surfaces consist of metal oxides, with the oxygen atoms in all cases prominently exposed at the surface of the mineral, some are very hydrophobic. As shown above (Table 8.3), hydrophilicity is mainly determined by a high value of the electron donor parameter ($\gamma^{\ominus}$) of the polar component (γ^{AB}) of the surface tension. This strong electron-donicity (Lewis basicity) is mainly due to the relatively loosely bound oxygen atoms at the surface of hydrophilic clay particles, such as smectites, vermiculites, micas or other metal oxide surfaces, such as silica (see Chapter 9 for the $\gamma^{\ominus}$-values of various clay and other mineral particles).

Talc and pyrophyllite on the other hand are quite hydrophobic, as seen from the high negative values of the free energies of interfacial interaction between their particles, when immersed in water (ΔG^{IF}_{1w1}); See Table 8.14. This is because talc and pyrophyllite are 2:1 layer silicate minerals with essentially no ionic substitutions and hence no layer charge. Thus, the external 001 oxygen atoms are completely satisfied in the sense of Pauling's electrostatic rule, being bonded to two tetrahedrally coordinated silicon atoms [1]; see also Giese *et al.* (1996). Therefore, in talc the basal oxygen atoms do not function as Lewis bases to any significant degree, in contrast with, e.g., smectites, vermiculites and micas.

[1]This is the sole and sufficient reason for the hydrophobicity of the surfaces of talc and pyrophyllite; see also Lewis (1923). That hydrophobicity is exceedingly unlikely to have any connection with the ease or otherwise with which, e.g., talc, is prone to adsorb nitrogen more or less strongly than other minerals (Michot *et al.*, 1990) nor is there any reason why a surface having adsorbed N_2, would therefore become more hydrophobic. As the physisorption of N_2 is almost exclusively driven by van der Waals attractions, one would expect that, e.g., smectites adsorb N_2 more strongly than talc, as they have a van der Waals constant (which is proportional to γ^{LW}) that is on average about 20% higher than talc, but nonetheless smectites are much more hydrophilic than talc; see Tables 9.1 and 9.2.

8.16 Special Properties of Kaolinite

Since the 17th Century in Western Europe, and much earlier in China and Japan, the clay mineral we know as kaolinite has been identified as the most appropriate raw material with which the finest porcelain ceramics ("China") can be made. Ceramic flatware, eating ware, statuary, etc., which is made from other clay minerals, tends to be coarser and is usually alluded to as earthenware; see Chapter 2.

Kaolinite particles differ in a few quantitatively measurable properties from other clay particles. Most clay and other mineral particles, when suspended in water, obey the classical laws of colloidal suspensions in that they tend to form more stable suspensions at low ionic strengths than at high ionic strengths (Verwey and Overbeek, 1948, 1999). Aqueous suspensions of kaolinite, however, display the exact opposite behavior; they are most stable in the presence of salt (NaCl), while they flocculate in distilled water (Schofield and Samson, 1954). The reason for this is that, like many species of platey clay particles, the flat sides of the particles are negatively charged, whilst their edges are positively charged. Thus, the negatively charged plates repel one another, but edges and plates attract each other. Now with most clay minerals, the edges are quite thin, so that the electrostatic repulsion between the plates by far outweighs the rather feeble attraction betweeen thin edges and much wider flat plates. In these (most common) cases, avoidance of flocculation is more strongly favored at low ionic strengths. With kaolinite, however, the (positively charged) edges are much thicker (typically of a thickness that is from 15 to 30% of the main dimension of the plates; see, e.g., Caillère *et al.* (1982), so that under appropriate conditions of low ionic strength of the suspending aqueous medium, attraction between edges and plates will predominate rather than repulsion between plates. It should be recalled that low ambient ionic strengths do not favor repulsion or attraction, they just favor electrical double layer interactions. Therefore, when one has mainly negatively charged platey surfaces, low ionic strengths promote the mutual *repulsion* between the plates, thus preventing flocculation. However, under conditions of close to equal numbers and sizes of plates and thick edges, which exist with kaolinite, the enhanced electrical double layer interaction energies occurring at low ionic strengths favor the *attraction* between oppositely charged (negative) plates and (positive) edges, causing the flocculation of kaolinite in a house-of-cards configuration. This then gives rise to an exceedingly stable and sturdy flocculated configuration, which is the underlying cause of the high quality ceramic, that is typical for porcelain, and which is only obtainable with kaolinite, house-of-cards like structures are also temporarlily obtainable with other types of clay (van Olphen, 1977), but these structures tend to be fleeting, on account of the thinness of their edges, and do not persist through the rather drastic flaming conditions required for the solidification of ceramics. Hence, kaolinite remains the most significant component for manufacturing real porcelain (van Olphen, 1977).

9

Surface Thermodynamic Properties of Minerals

One of the great advantages of studying phyllosilicate minerals lies in the fact that the structure of the major external faces, the 001 surfaces, is well known. This arises from our knowledge of the atomic arrangement of e.g., micas and related minerals. The perfect 001 cleavage ensures that the external surface is populated by a known arrangement of atoms. The caveats in this are the uncertainty of the placement of the interlayer cations. When muscovite mica is cleaved, the interlayer potassium ions (see Chapter 3) must distribute themselves between the two new surfaces. Presumably, about half the potassium ions go with one new surface and the rest with the other. The exact positions of the potassium on the surfaces are not known.

9.1 Phyllosilicate Minerals

9.1.1 Samples

The Clay Minerals Society offers a repository of standard clay minerals which are available for a modest charge. Other clay minerals, such as the BP Colloid, were obtained from industrial sources. The BP Colloid was supplied by Southern Clay Products, Gonzales, TX, Polar Gel, Volclay and Panther Creek were obtained from American Colloid Co., Skokie, IL. As is usual with clay minerals, the <2 μm fraction was isolated by centrifugation of aqueous dispersions of the clay samples. The wet clays were dried by mild heating.

To obtain samples exchanged with specific cations, 2 grams of clay sample were dispersed in 50 ml of a 0.1 M chloride solution and stirred for two days. Subsequently, the clay was removed by centrifugation; the procedure was performed a total of three times. Excess salt was then removed from the treated clay by dialysis. The exchanged clays were dried by mild heat.

The contact angles for the clay samples were determined by direct measurement of drops of test liquids on oriented films of the clays. Some clay minerals are capable of forming self supporting films, but this was not true of all the clay minerals listed here. In order to have comparable measurements for the all the clay minerals, it was necessary to use the same preparative technique for all. One technique which worked for all the minerals was to sediment an aqueous suspension of the clay onto clean glass slides and allow the water to evaporate overnight. The films were dried at 110°C for 1 hour to remove pore water, and then stored in a desiccator until needed. The samples were finally allowed to equilibrate with the atmosphere before measurement.

For each sample, i.e., a clay mineral saturated with a specific cation or untreated, contact angle measurement were performed as described in Chapter 7 using five liquids; α-bromonaphthalene, diiodomethane, water, formamide and glycerol. The contact angles were used to solve the simultaneous Young equations to yield the values of γ^{LW}, $\gamma^{\oplus}$, and $\gamma^{\ominus}$.

9.1.2 Values

The surface tension components and parameters are listed in Table 9.1 along with the calculated values for the acid-base part of the surface tension, γ^{AB} and the total surface tension for each clay sample, γ^{Tot}. In addition, the interaction energies, ΔG^{IF}_{1w1} and their contributors, ΔG^{LW}_{1w1} and ΔG^{AB}_{1w1} are also listed, all in mJ/m^2, for clay particles immersed in water. Positive values of ΔG^{IF}_{1w1} show that the clay surfaces are hydrophilic while negative values correspond to hydrophobic particles (van Oss and Giese, 1995).

Clay	Cation	γ^{LW}	$\gamma^{\oplus}$	$\gamma^{\ominus}$	γ^{AB}	γ^{Tot}	ΔG^{LW}_{1w1}	ΔG^{AB}_{1w1}	ΔG^{IF}_{1w1}
SAz-1: cec = 120 meq/100g									
	K	43.1	1.1	39.0	13.3	56.5	-7.2	19.0	11.83
	Na	42.4	2.3	33.4	17.3	59.8	-6.8	10.4	3.56
	Amm	40.9	1.7	37.0	15.8	56.7	-6.0	15.5	9.48
	Cs	46.4	1.0	46.6	13.5	59.9	-9.2	28.9	19.78
	Li	42.2	2.0	35.7	16.7	58.9	-6.6	13.5	6.82
	Mg	43.4	1.8	39.3	16.8	60.2	-7.3	18.1	10.74
	Ca	42.2	1.6	42.7	16.7	58.9	-6.7	22.4	15.68
	Ba	42.1	1.5	29.0	13.2	55.3	-6.6	5.1	-1.49

Clay	Cation	γ^{LW}	$\gamma^{\oplus}$	$\gamma^{\ominus}$	γ^{AB}	γ^{Tot}	ΔG^{LW}_{1w1}	ΔG^{AB}_{1w1}	ΔG^{IF}_{1w1}
	Sr	43.4	1.4	44.4	15.7	59.1	-7.4	25.1	17.69
	Nat	43.7	1.4	46.9	16.4	60.0	-7.5	27.8	20.24
SWa-1: cec = 98 meq/100g									
	K	43.6	1.9	36.2	16.8	60.4	-7.5	14.1	6.61
	Na	43.2	2.3	32.5	17.5	60.7	-7.3	9.2	1.96
	Amm	40.7	1.8	35.6	16.0	56.7	-5.9	13.6	7.75
	Cs	50.2	1.5	38.8	15.0	65.2	-11.7	18.1	6.42
	Li	42.0	2.9	25.9	17.3	59.2	-6.5	0.5	-6.00
	Mg	44.4	1.6	41.6	16.2	60.6	-7.9	21.2	13.26
	Ca	44.1	1.5	40.1	15.4	59.4	-7.8	19.7	11.88
	Ba	42.4	1.5	37.2	15.1	57.5	-6.8	16.0	9.18
	Sr	42.6	1.7	36.0	15.7	58.3	-6.9	14.2	7.32
	Nat	43.6	1.8	36.8	16.2	59.8	-7.5	15.0	7.56
Panther Creek: cec = 88 meq/100g									
	K	43.1	1.2	40.7	14.0	57.1	-7.2	21.0	13.76
	Na	43.5	1.3	33.1	13.3	56.8	-7.4	11.0	3.58
	Amm	42.1	0.9	38.3	12.0	54.1	-6.6	18.6	11.93
	Cs	46.7	1.2	46.2	15.0	61.6	-9.4	27.6	18.27
	Li	42.2	1.8	33.9	15.6	57.8	-6.7	11.4	4.76
	Mg	43.9	1.5	40.3	15.8	59.6	-7.6	19.8	12.18
	Ca	44.5	1.2	42.0	14.3	58.8	-8.0	22.6	14.62
	Ba	43.2	1.0	40.7	12.4	55.6	-7.2	21.7	14.44
	Sr	43.7	1.2	44.1	14.5	58.2	-7.5	25.2	17.63
	Nat	44.7	1.5	44.8	16.2	60.9	-8.1	25.2	17.04
Bentolite-L: cec = 88 meq/100g									
	K	43.7	0.3	47.9	6.9	50.5	-7.5	34.0	26.53
	Na	42.9	1.3	33.6	13.4	56.3	-7.1	11.6	4.52

Clay	Cation	γ^{LW}	$\gamma^{\oplus}$	$\gamma^{\ominus}$	γ^{AB}	γ^{Tot}	ΔG^{LW}_{1w1}	ΔG^{AB}_{1w1}	ΔG^{IF}_{1w1}
	Amm	42.5	1.1	44.0	13.7	56.1	-6.8	25.4	18.57
	Cs	46.0	1.3	41.6	14.7	60.6	-8.9	22.0	13.05
	Li	43.0	2.2	34.5	17.5	60.5	-7.2	11.7	4.56
	Mg	44.2	1.3	44.1	15.1	59.3	-7.9	24.9	17.08
	Ca	44.4	1.2	44.0	14.7	59.2	-8.0	24.9	16.95
	Ba	43.8	1.3	42.6	14.7	58.5	-7.6	23.2	15.59
	Sr	43.6	0.9	40.6	11.9	55.5	-7.5	21.8	14.32
	Nat	44.5	1.4	45.2	16.7	60.1	-8.0	26.0	18.00
STx-1:cec = 80 meq/100g									
	K	43.2	1.1	40.9	13.5	46.7	-7.2	21.6	14.32
	Na	43.2	2.1	33.1	16.7	59.9	-7.2	10.1	2.91
	Amm	42.2	1.4	43.1	15.8	57.9	-6.7	23.4	16.68
	Cs	46.3	1.1	42.5	13.7	60.0	-9.1	23.5	14.35
	Li	42.9	2.0	37.9	17.4	60.3	-7.1	16.1	9.03
	Mg	43.9	1.5	43.3	16.1	60.0	-7.7	23.4	15.77
	Ca	44.3	1.3	44.5	15.1	59.3	-7.9	25.4	17.55
	Ba	44.3	1.3	43.4	15.0	59.3	-7.9	24.0	16.11
	Sr	44.0	1.1	42.4	13.5	57.5	-7.7	23.5	15.79
	Nat	44.0	1.2	45.8	15.0	59.0	-7.7	27.1	19.39
Volclay: cec = 80 meq/100g									
	K	42.1	1.5	28.9	13.1	55.2	-6.6	5.0	-1.62
	Na	42.4	1.3	28.0	12.1	54.5	-6.8	3.8	-2.98
	Amm	41.4	2.2	31.2	16.6	57.9	-6.2	7.6	1.41
	Cs	42.5	1.7	30.5	14.2	56.7	-6.8	7.1	0.27
	Li	41.5	3.7	17.1	15.8	57.3	-6.3	-11.5	-17.75
	Mg	41.9	1.4	39.2	14.9	56.7	-6.5	18.7	12.20
	Ca	42.8	1.4	39.0	14.8	57.7	-7.0	18.5	11.43

Clay	Cation	γ^{LW}	$\gamma^{\oplus}$	$\gamma^{\ominus}$	γ^{AB}	γ^{Tot}	ΔG^{LW}_{1w1}	ΔG^{AB}_{1w1}	ΔG^{IF}_{1w1}
	Ba	42.2	1.4	35.5	13.8	56.0	-6.6	14.2	7.52
	Sr	43.4	1.5	37.4	14.8	58.1	-7.3	16.4	9.05
	Nat	42.4	1.5	34.7	14.6	56.9	-6.8	12.8	6.06
Hectorite: cec = 78 meq/100g									
	K	43.6	1.5	39.1	15.5	59.1	-7.5	18.4	10.92
	Na	43.3	1.5	39.8	15.5	58.8	-7.3	19.3	11.98
	Amm	41.6	1.5	39.8	15.6	56.9	-6.3	19.2	12.87
	Cs	43.9	1.5	42.3	15.7	59.5	-7.6	22.4	14.76
	Li	42.5	1.9	40.6	17.3	59.9	-6.9	19.5	12.68
	Mg	43.4	1.5	39.9	15.7	59.0	-7.3	19.3	12.00
	Ca	43.6	1.4	40.6	14.8	58.4	-7.5	20.5	13.07
	Ba	41.4	1.4	40.3	14.8	56.1	6.2	20.1	13.91
	Sr	42.3	1.8	40.0	17.0	59.3	-6.7	18.9	12.20
	Nat	42.5	2.2	36.6	17.7	60.2	-6.8	14.4	7.55
BP Colloid: cec = 75 meq/100g									
	K	42.4	1.1	31.3	11.7	54.1	-6.8	8.6	1.89
	Na	42.4	1.8	31.8	15.1	57.5	-6.8	8.7	1.93
	Amm	41.5	1.9	33.7	15.9	57.4	-6.3	11.1	4.81
	Cs	42.5	2.3	26.2	15.5	57.9	-6.8	1.0	-5.79
	Li	41.5	3.2	28.5	19.2	60.6	-6.3	3.8	-2.45
	Mg	43.1	1.5	40.0	15.3	58.4	-7.2	19.6	12.44
	Ca	43.1	1.3	38.9	13.9	57.0	-7.2	18.7	11.51
	Ba	42.2	1.8	35.1	15.9	58.1	-6.7	13.0	6.36
	Sr	41.5	2.2	38.8	18.6	60.1	-6.3	16.8	10.53
	Nat	41.3	2.6	29.4	17.5	58.8	-6.2	5.2	-1.02
Saponite (SapCa-1): cec = 68 meq/100g									
	K	40.8	3.3	16.8	14.8	55.6	-5.9	-12.4	-18.29

Clay	Cation	γ^{LW}	$\gamma^{\oplus}$	$\gamma^{\ominus}$	γ^{AB}	γ^{Tot}	ΔG^{LW}_{1w1}	ΔG^{AB}_{1w1}	ΔG^{IF}_{1w1}
	Na	43.0	2.3	29.0	16.5	59.4	-7.1	4.7	-2.39
	Amm	41.1	1.7	30.5	14.6	55.7	-6.1	7.1	1.03
	Cs	42.0	2.5	17.0	13.2	55.2	-6.6	-12.7	-19.34
	Li	41.5	2.3	32.7	17.2	58.7	-6.3	9.4	3.16
	Mg	42.9	1.9	37.5	17.0	59.9	-7.1	15.7	8.63
	Ca	43.3	1.6	41.4	16.3	59.6	-7.3	20.9	13.59
	Ba	40.0	2.3	25.6	15.2	55.2	-5.5	0.2	-5.32
	Sr	42.5	1.8	39.0	16.9	59.4	-6.8	17.6	10.78
	Nat	41.2	2.6	28.1	17.1	58.2	-6.1	3.4	-2.68
SWy-1: cec = 68 meq/100g									
	K	42.1	1.1	30.5	11.4	53.5	-6.6	7.6	1.02
	Na	43.0	1.5	36.7	14.8	57.7	-7.1	15.4	8.30
	Amm	41.7	0.7	36.2	10.3	52.0	-6.4	16.2	9.80
	Cs	42.5	1.9	27.4	14.4	56.9	-6.8	2.7	-4.14
	Li	42.3	2.7	31.1	18.5	60.8	-6.7	7.2	0.48
	Mg	42.9	0.9	43.3	12.5	55.4	-7.1	25.1	18.00
	Ca	42.5	1.2	41.1	14.1	56.7	-6.9	21.5	14.65
	Ba	40.2	2.0	20.2	12.6	52.8	-5.6	-8.1	-13.74
	Sr	41.4	1.5	30.8	13.5	54.9	-6.2	7.7	1.43
	Nat	41.2	1.5	33.3	14.3	55.5	-6.1	11.0	4.83
Polar Gel: cec = 60 meq/100g									
	K	44.0	1.2	43.5	14.4	58.4	-7.7	24.5	16.81
	Na	43.7	1.5	41.3	15.8	59.5	-7.5	21.0	13.48
	Amm	42.5	1.8	43.7	17.5	60.0	-6.8	23.3	16.46
	Cs	44.7	1.2	44.9	14.7	59.4	-8.1	26.1	17.85
	Li	42.1	2.0	42.3	18.6	60.7	-6.6	21.1	14.45
	Mg	44.2	1.4	44.2	15.5	59.7	-7.8	24.8	16.99

Clay	Cation	γ^{LW}	$\gamma^{\oplus}$	$\gamma^{\ominus}$	γ^{AB}	γ^{Tot}	ΔG^{LW}_{1w1}	ΔG^{AB}_{1w1}	ΔG^{IF}_{1w1}
	Ca	44.5	1.3	43.4	15.1	59.6	-8.0	24.0	15.98
	Ba	44.3	1.0	46.4	13.7	57.9	-7.9	28.5	20.62
	Sr	44.8	1.3	45.9	15.5	60.3	-8.2	27.0	18.78
	Nat	43.9	1.7	45.1	17.3	61.3	-7.7	25.0	17.32
Laponite RD: cec = 115 meq/100g									
	K	43.2	1.4	31.8	13.3	56.5	-7.2	9.1	1.87
	Na	42.3	2.1	29.8	15.8	58.1	-6.7	5.9	-0.84
	Amm	42.5	1.5	30.7	13.6	56.1	-6.8	7.5	0.67
	Cs	43.9	1.5	37.0	14.9	58.8	-7.7	15.8	8.15
	Li	43.1	1.5	34.7	14.4	57.5	-7.2	12.9	5.68
	Mg	43.1	1.5	31.6	13.8	56.9	-7.2	8.7	1.56
	Ca	42.9	1.4	30.6	12.0	55.8	-7.1	6.0	-1.03
	Ba	42.6	2.0	28.8	15.2	57.8	-6.9	4.6	-2.30
	Sr	42.3	1.6	30.1	13.9	56.2	-6.7	6.6	-0.12
	Rb	43.9	1.5	37.0	14.9	58.8	-7.7	15.8	8.15
	Nat	41.3	3.0	24.2	17.0	58.3	-6.2	-1.7	-7.91
	Nat[a]	44.4	1.3	40.5	14.5	58.9	-8.0	20.6	12.60

Table 9.1: *Surface tension values and parameters for a group of clay minerals, each saturated with a specific cation (Norris, 1993; Norris et al., 1993). The heading "Nat" refers to the clay with no cation exchange treatment. "Amm" refers to an exchange with the ammonium cation. All values are in units of mJ/m² measured at 20° C. The values listed under ΔG^{IF}_{1w1} are the calculated interaction energies, in mJ/m², for the particular clay dispersed in water. The Nat[a] sample of Laponite RD was dialysed against distilled water before measurement.*

9.1.3 Generalities; clay minerals

All the minerals listed in Table 9.1 are smectites. They differ in their chemistry and where the layer charge resides. Examination of Table 9.1 shows several general patterns in the values of the various surface thermodynamic components and parameters. The values of γ^{LW} are fairly constant, varying between 40.0 and 50.2 mJ/m². Note that the maximum value of 50.2

mJ/m^2, for the Cs saturated SWa-1 clay, is an unusually large value (see Figure 9.1). The average for γ^{LW} is 43.0 mJ/m^2 with a standard deviation of 1.4 mJ/m^2. In contrast, the Lewis acid parameter, $\gamma^{\oplus}$, is uniformely small varying between 0.3 and 3.7 mJ/m^2 with an average of 1.6 mJ/m^2 and a standard deviation of 0.5 mJ/m^2. This is in general agreement with many measurements of condensed media (van Oss 1994). As has been pointed out elsewhere, the major variable in the surface thermodynamic properties of condensed materials is the value of the Lewis base paramter, $\gamma^{\ominus}$. This is certainly true for the clay samples listed in Table 9.1. These values vary between 16.8 and 47.9 mJ/m^2 with an average value of 36.9 mJ/m^2 and a large standard deviation of 6.7 mJ/m^2. These trends are shown in Figure 9.1 where the values of γ^{LW} are plotted against the values of $\gamma^{\ominus}$. There is an apparent modest inverse relation between γ^{LW} and $\gamma^{\ominus}$. The origin of this relation is not immediately obvious.

As discussed earlier, the approximate boundary between hydrophobic materials and hydrophilic materials occurs for $\gamma^{\ominus} \approx 28$ mJ/m^2. This boundary is shown as a horizontal line in Figure 9.1. In addition to the relation shown in Figure 9.1, there is also a modest inverse relation between $\gamma^{\ominus}$ and $\gamma^{\oplus}$ (Figure 9.2) for the clay minerals listed in Table 9.1. The origins of the Lewis acid and base interactions are directly related to the ability of a material to donate or accept electrons at an interface. For example, the most obvious manifestation of these interactions is hydrogen bonding where an electronegative atom or atoms act as Lewis bases (e.g., oxygen) and an electropositive atom or atoms act as Lewis acids (e.g., hydrogen in a water molecule or a Ca^{2+} ion). Normally one would expect these two functionalities, when found on the same material , e.g., a clay mineral surface, to be unrelated. These trends can also be seen in the surface tension values of specific clay minerals. For example, the iron-rich clay, SWa-1, shows a nearly linear inverse relation between $\gamma^{\ominus}$ and $\gamma^{\oplus}$ with the monovalent interlayer cations arranged toward the smaller $\gamma^{\ominus}$ values (except Cs) and the divalent interlayer cations falling at larger values of $\gamma^{\ominus}$ (Figure 9.3). Not all of the clays in question exhibit such a clear linear relation. For example, plotting $\gamma^{\ominus}$ versus $\gamma^{\oplus}$ for the smectite clay mineral, STx-1, shows all the cations except Na and Li clustered at large $\gamma^{\ominus}$ and small $\gamma^{\oplus}$-values (Figure 9.4). There is still an inverse relation but that trend is determined by just two of the cations.

For the natural clay minerals in Table 9.1 (i.e., excluding Laponite which is a synthetic clay), the predominance of the clay samples with large $\gamma^{\oplus}$ values (taken arbitrarily as $\gamma^{\oplus} >2$ mJ/m^2) tend to be lithium-saturated clays followed in abundance by cesium-saturated clays. It has been known for some

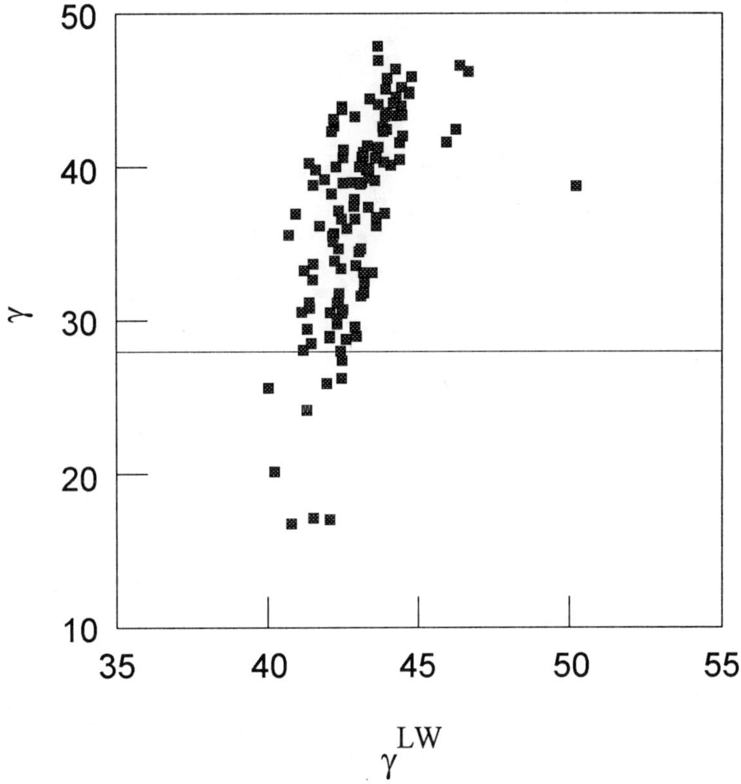

Figure 9.1: *A plot of* γ^{LW} *versus* $\gamma^{\ominus}$ *for the clay samples listed in Table 9.1. The horizontal line represents the approximate boundary between hydrophilic clay samples (above the line) and hydrophobic clay samples (below the line).*

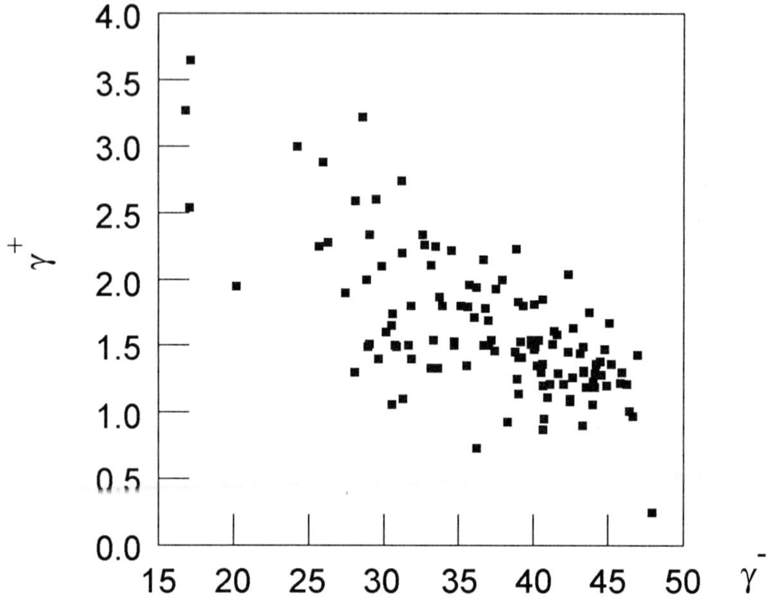

Figure 9.2: *A plot of* $\gamma^{\oplus}$ *versus* $\gamma^{\ominus}$ *for the clay samples listed in Table 9.1.*

time that lithium can play an unusual role in clay mineral structures, e.g., the diffusion of interlayer lithium cations into the vacant octahedral sites of a dioctahedral smectite under mild heating.

It can be seen (Table 9.1) by the ΔG^{IF}_{1w1} values that the large majority of smectites are hydrophilic, in the presence of most cations (i.e., $\Delta G^{IF}_{1w1} > 0$). However, in a few cases some alkali metal counterions cause hydrophobicity: e.g., Volclay (K^+, Na^+, Li^+); BP Mineral Colloid (Cs^+, Li^+), saponite (K^+, Na^+, Cs^+); SWy-1 (Cs^+) and Laponite RD (Na^+). Also in a few cases alkali earth cations accompany hydrophobicity: e.g., SAz-1 (Ba^{2+}); SWy-1 (Ba^{2+}); Laponite RD (Ca^{2+}, Sr^{2+}). In such cases the (negative) ζ-potentials of these particles are prone to be lowest when bound to these cations (Wu *et al.*, 1994a).

As discussed earlier, the interlayer cations and those which are adsorbed to the external surfaces of smectite clay minerals are hydrated. The nature of the hydration shell should be related to the ionic charge of the cation;

Amine Coverage (wt %)	γ^{LW} mJ/m^2	$\gamma^{\oplus}$ mJ/m^2	$\gamma^{\ominus}$ mJ/m^2
0	28.5	0.1	11.6
0.5	23.5	0	2.0
1.0	19.3	0	0
1.5	17.4	0	0
2.0	16.4	0	0
2.5	16.0	0	0
3.0	15.8	0	0

Table 9.2: *Talc powders which were treated with varying amounts of octadecylamine with mild heating. As the coverage of the talc by the amine increases, the polar properties of the talc are quickly lost, and the LW value also decreases (Li et al., 1993).*

the monovalent cations should be less strongly hydrated than the divalent cations. This difference should be reflected in the values of the surface tension components and parameters for the different cation-saturated clay minerals in Table 9.1. Plotting γ^{LW} against the number of electrons in the different cations seems to show such a difference (Figure 9.5). Here, for the divalent cations, γ^{LW} decreases with increasing number of electrons while the monovalent cations behave in the opposite sense. Such divergent behavior is observed for almost all the smectite clay minerals in Table 9.1 with the exception of the commercial clay Volclay. This has been interpreted to reflect the fact that the monovalent cations are likely in direct contact with the clay surface with the hydrating water molecules in contact with the exposed part of the cation. For the divalent cations, the higher hydration energy keeps water around all sides of the cation, and thus the divalent cation does not physically touch the clay surface; there is always water between the two (Norris, 1993).

9.1.4 Role of organic material adsorbed on clays

Organic material can be positioned on a clay mineral surface in two ways; either by the exchange of the native inorganic cations by an organic cation (e.g., a quaternary ammonium cation) or by surface adsorption of a neutral organic molecule (e.g., an amine) on the external surface of the clay particles. Tables 9.2 and 9.3 illustrate the changes in surface thermodynamic properties as the coverage of the clay surfaces varies. It is clear that modest amounts

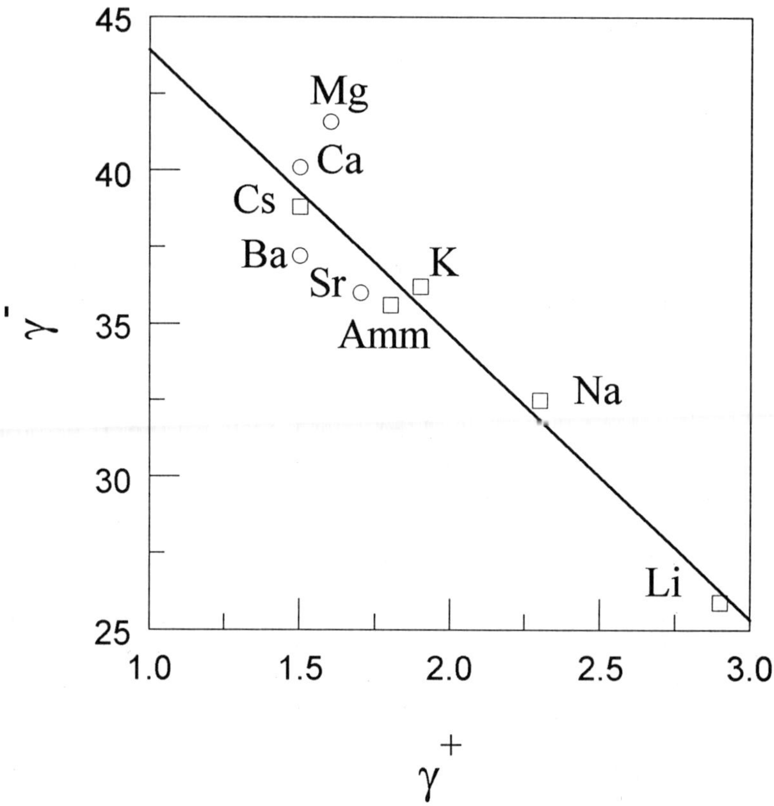

Figure 9.3: *A plot of* $\gamma^{\oplus}$ *versus* $\gamma^{\ominus}$ *for the smectite clay SWa-1. The line is a least squares fit to all the data points.*

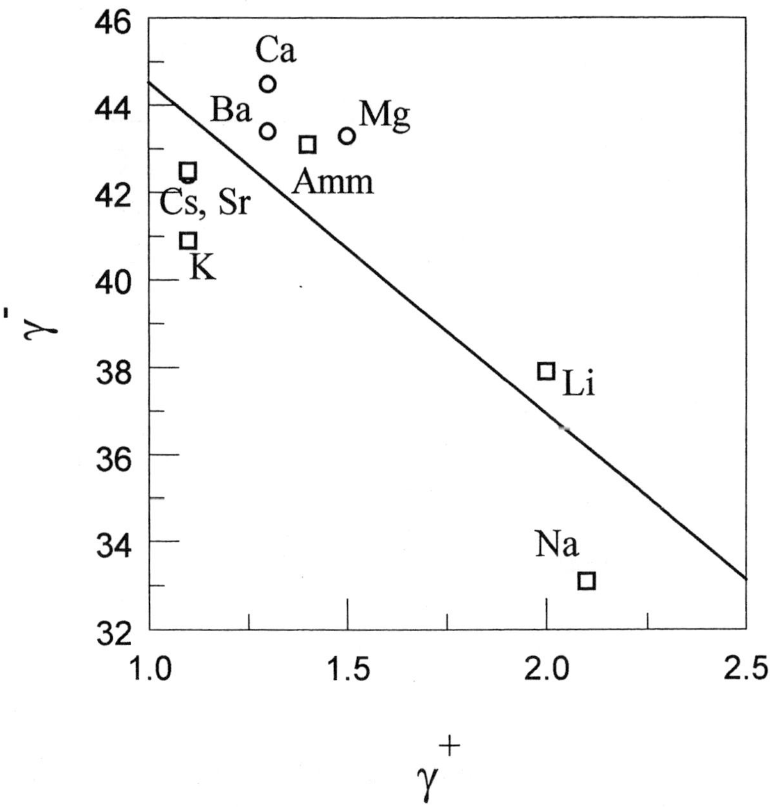

Figure 9.4: *A plot of* $\gamma^{\oplus}$ *versus* $\gamma^{\ominus}$ *for the smectite clay STx-1. The line is a least squares fit to all the data points.*

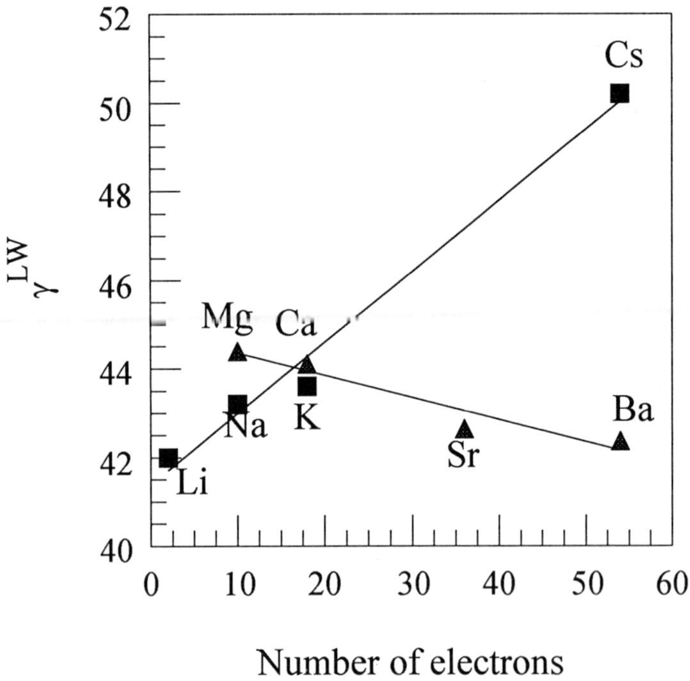

Figure 9.5: *A plot of* γ^{LW} *versus the number of electrons in the exchangeable cation for the smectite clay SWa-1. The lines are a least squares fit to all the monovalent and divalent cation data points.*

of organic matter, either adsorbed or bonded to the clay mineral surfaces, renders the composite material more hydrophobic. This has been exploited in a number applications ranging from lubricants to barriers surrounding toxic dump sites.

Mineral	γ^{LW}	$\gamma^{\oplus}$	$\gamma^{\ominus}$	γ^{AB}	γ^{Tot}	ΔG_{1w1}^{IF}
Wyoming montmorillonite (SWy-1)						
Number of carbon atoms						
0	41.7	0.7	36.2	10.1	51.8	9.9
6	41.2	0.7	9.8	5.2	46.4	-38.5
7	39.9	1.0	8.4	5.8	45.7	-40.3
8	39.4	1.0	7.7	5.5	44.9	-42.0
9	39.1	0.7	7.5	4.6	43.7	-44.0
10	40.5	0.3	15.8	1.1	41.0	25.1
11	40.5	0.3	7.5	3.0	43.5	-47.4
12	39.6	0.3	6.0	2.7	42.3	-52.1
13	40.4	0.2	7.2	2.4	72.8	-71.3
14	39.5	0.4	5.3	2.9	42.4	-53.8
15	37.7	0.3	0.1	0.3	38.0	-89.6
Other organic cations						
trimethyl ammonium	41.5	3.0	12.9	12.4	53.9	-25.6
hexadecyl trimethyl ammonium	40.0	0.5	8.7	4.2	44.2	-42.0
trimethyl phenyl ammonium	41.3	3.1	9.7	11.0	52.3	-31.6
tetraethyl ammonium	41.3	1.5	19.5	10.8	52.1	-15.9
Laponite RD						
Number of carbon atoms						
0	42.5	1.5	30.7	13.6	56.1	0.7
8	42.1	1.0	24.2	9.8	51.9	-8.7
10	41.4	0.8	24.3	8.8	50.2	-8.2
12	41.5	0.9	24.5	9.4	50.9	-7.9
Other organic cations						

Mineral	γ^{LW}	$\gamma^{\oplus}$	$\gamma^{\ominus}$	γ^{AB}	γ^{Tot}	ΔG^{IF}_{1w1}
hexadecyl trimethyl ammonium	43.0	0.6	24.7	7.7	50.7	-8.5

Table 9.3: *Surface tension components and parameters for Wyoming montmorillonite and Laponite whose exchangeable cations have been exchanged by n-alkyl ammonium cations. All values are in units of mJ/m² measured at 20° C.*

9.2 Other Minerals

Table 9.4 lists values of the surface thermodynamic components and parameters for a number of minerals. Clearly, these values are not intended to be anything more than a sampling of the mineral kingdom. The choices were driven partly by the importance of the minerals in the earth's crust (e.g., calcite, quartz), partly by the importance of the mineral in technological applications (e.g., zeolites) and partly to examine a wide range of mineral types and chemistries. It is hoped that further work will build on these preliminary observations.

Mineral	γ^{LW}	$\gamma^{\oplus}$	$\gamma^{\ominus}$	γ^{AB}	γ^{Tot}	ΔG^{IF}_{1w1}
Micas						
Margarite, Chester MA	38.1	0.4	57.6	9.6	47.7	40.4
Phlogopite, Madagascar	40.8	0.6	59.3	11.9	52.7	39.4
Phlogopite,Ontario	39.4	1.4	54.5	17.5	56.9	30.9
Muscovite,Keystone, S.D.	40.6	1.8	51.5	19.3	59.9	25.7
Biotite, Bancroft, Ontario	42.0	1.4	47.8	16.4	58.4	22.3
Carbonates						
Calcite, Creel, Mexico	40.2	1.3	54.4	16.8	57.0	30.8
Dolomite, Barstow, CA	37.6	0.2	30.5	4.9	42.5	4.4
Magnesite, Baumado Bahia, Brazil	30.0	0.0	7.8	0.0	30.3	-48.1
Vermiculite						
Vermiculite, Kellogg, ID	38.4	1.0	59.7	15.5	53.9	38.7
Zeolites						
Valfor 100 (zeolite 4A)	28.1	0.7	56.2	12.5	40.6	40.4

Mineral	γ^{LW}	$\gamma^{\oplus}$	$\gamma^{\ominus}$	γ^{AB}	γ^{Tot}	ΔG^{IF}_{1w1}
Purmol 13X (zeolite 13X)	30.4	2.0	43.4	18.6	49.0	20.9
Ethyl 4A	28.3	2.2	24.7	14.7	43.0	-2.0
Erionite	36.1	0.2	12.7	3.2	39.3	-30.9
Oxides						
Fe_2O_3 (synthetic hematite)	45.6	0.3	50.4	7.8	53.4	28.2
Hematite, Urals	36.5	0.0	17.8	0.0	36.5	-20.5
Magnetite, Essex, NY	42.2	0.0	24.4	0.0	42.2	-8.9
TiO_2 (Fisher Scientific)	42.1	0.6	46.3	10.5	52.6	23.4
Rutile, York Co, PA	40.8	0.3	32.8	6.3	47.1	6.3
SiO_2 (Fisher scientific)	39.2	0.8	41.4	11.5	50.7	17.9
α Al_2O_3 (Aldrich)	31.6	0.6	27.2	8.1	39.7	1.0
ZrO_2 (Matthey, MA)	34.8	1.3	3.6	4.3	39.1	-52.3
SnO_2 (Matthey, MA)	31.1	2.9	8.5	9.9	41.0	-30.2
Phosphate						
Apatite, Durango, Mexico	35.4	0.0	20.5	0.0	35.4	-13.8
Fluor-apatite, synthetic	32.4	0.6	9.0	4.6	37.0	-37.1
Hydroxy-apatite, synthetic	36.2	0.9	16.0	7.6	43.8	-20.9
Sorosilicate						
Vesuvianite, Chihuahua, Mexico	38.1	0.0	49.8	0.0	38.1	36.0
Sulphate						
Anhydrite, Gabbs, Nevada	39.2	1.2	48.8	15.3	54.5	25.6
Barite, Sterling, Colorado	41.0	0.1	40.6	4.0	45.0	19.0
Celestite, Ontario	32.3	0.8	16.0	7.2	39.5	-19.5
Gypsum, St. George, Utah	50.0	1.4	42.8	15.5	65.5	11.5
Tectosilicate						
Albite, Ontario	37.0	0.4	22.5	6.0	43.0	-9.4
Andesine, Norway	40.8	0.0	32.2	0.0	40.8	6.7

Mineral	γ^{LW}	$\gamma^{\oplus}$	$\gamma^{\ominus}$	γ^{AB}	γ^{Tot}	ΔG^{IF}_{1w1}
Orthoclase, Bancroft, Ontario	34.0	0.6	21.8	7.2	41.2	-9.2
Oligoclase, Sanimal, Norway	42.0	0.0	30.5	0.0	42.0	3.0
Quartz, Hot Springs, Arkansas	30.3	0.1	37.4	3.9	34.2	18.8
Illite						
DP-10	37.5	1.6	46.6	17.3	54.8	22.7
LSH	40.2	1.3	42.6	14.9	55.1	17.5
93-6-10A	39.0	1.8	43.0	17.6	56.6	17.4
Kaolinite						
Kaolinite (KGa-1)	35.9	0.4	34.4	7.4	43.3	10.9
Synthetic Hydrotalcites						
H-3 (Al/Mg = 0.43)	21.1	0.0	26.8	0.0	21.1	2.6
H-4 (Al/Mg = 0.25)	24.6	0.0	31.1	0.0	24.0	10.5
H-5 (Al/Mg = 0.33)	27.0	0.0	18.0	0.0	27.0	-16.9
Asbestos minerals						
Chrysotile						
Globe, AZ	35.1	0.0	31.2	0.0	35.1	7.7
Ontario	40.4	0.5	25.8	7.2	47.6	-5.2
Jacob's Mine, Quebec	38.2	0.0	23.9	0.0	38.2	-7.8
Zimbabwe	37.7	0.0	6.2	0.0	37.7	-56.0
Carrey Mine, Quebec	35.2	2.9	15.4	13.4	48.6	-18.3
Salt River, AZ	42.7	0.0	5.0	0.0	42.7	-63.8
Asbestiform Amphiboles						
Crocidolite, Capetown, SA	37.4	0.9	23.9	9.3	46.7	-6.8
Crocidolite, Pziesk, SA	40.0	0.6	20.1	6.9	46.9	-15.2
Non-asbestiform Amphiboles						
Actinolite M41241	36.4	0.6	9.3	4.7	41.1	-37.9
Tremolite M19605	35.3	0.0	10.5	0.0	35.3	-39.8

Mineral	γ^{LW}	$\gamma^{\oplus}$	$\gamma^{\ominus}$	γ^{AB}	γ^{Tot}	ΔG^{IF}_{1w1}
Tremolite M36910	38.7	0.0	8.1	0.0	38.7	-49.3
Anthophyllite, Kagero, Norway	30.7	1.5	41.4	15.8	46.5	19.7
Enstatite, Bamble, Norway	39.9	0.0	11.9	0.0	39.9	-37.7
Spodumene, Brazil	39.1	0.4	19.1	5.5	44.6	-17.0
Cyclosilicate						
Schorl, S. Dakota	32.9	2.3	40.2	19.2	52.1	16.0
Uvite, Pierrepont, NY	38.1	0.1	26.0	3.2	41.3	-3.6
Halide						
Fluorite, Rosiclare, Ill.	26.6	0.4	1.5	1.5	28.1	-68.1
Hydroxide						
Brucite, Gabbs, Nevada	37.0	0.8	16.6	7.3	44.3	-20.2
Nesosilicate						
Almandine, Gore Mt., NY	33.2	0.0	12.6	0.0	33.2	-32.7
Almandine, Wrangell, Alaska	29.2	1.0	16.9	8.2	37.4	-16.3
Fayalite, El Paso Co, CO	36.7	0.0	6.5	0.0	36.7	-54.4
Grossular, Chihuahua, Mexico	35.9	0.2	21.1	4.1	40.0	-11.9
Pseudo-layer Structure Minerals						
Palygorskite PFL-1	29.5	0.0	28.7	0.0	29.5	5.0
Palygorskite SE China	29.2	0.0	13.4	0.0	29.2	-29.1
Sepiolite, Sp-1	30.5	0.1	22.5	3.0	33.5	-7.3
Sepiolite, Nevada	30.5	0.1	17.3	2.6	33.1	-18.3
Zero Layer Charge Phyllosilicates						
Talc (Fisher)	30.7	1.8	5.9	6.5	37.2	-40.4
Talc (Fisher, dried at 400°C)	34.2	0.1	6.9	1.7	35.9	-48.7
Talc, Gouverneur	31.5	2.4	2.7	5.1	36.6	-49.5
Pyrophyllite	33.9	1.7	4.9	5.8	39.7	-45.2
Other Materials						

Mineral	γ^{LW}	$\gamma^{\oplus}$	$\gamma^{\ominus}$	γ^{AB}	γ^{Tot}	ΔG^{IF}_{1w1}
Activated charcoal CX 0655-1	20.9	0.7	31.7	9.4	30.3	9.8
Ice	29.6	14.0	28.0	39.6	69.2	0.07
Solid Surfaces vs. Ground Powders						
Glass (microscope slide, VWR)	33.7	1.3	62.8	18.1	51.8	42.4
same glass, ground	31.1	0.4	37.1	7.7	38.8	16.7
Calcite, Chihuahua, Mexico	40.2	1.3	54.4	16.8	57.0	30.8
same calcite, ground	29.1	0.5	31.6	7.9	37.0	8.9
Dolomite, Barstow CA	37.6	0.2	30.5	4.9	42.5	4.4
same dolomite, ground	27.1	0.2	13.6	3.3	30.4	-25.6
Talc, massive (soapstone)	40.4	0.9	8.9	5.7	46.1	-39.6
same talc, ground	37.4	2.0	2.2	4.2	41.6	-56.0
Coarse vs. finely ground powders						
Coarse silica (see above)	39.2	0.8	41.4	11.5	50.7	17.9
fine silica	36.8	2.3	14.5	11.5	48.3	-21.5
Coarse zirconia (see above)	34.8	1.3	3.6	4.3	39.1	-52.3
fine zirconia	32.2	3.4	1.5	4.5	36.7	-51.1

Table 9.4: *Surface tension components and parameters for other clay minerals and non-clay minerals. All values are in units of mJ/m^2 measured at 20°C. The values listed under ΔG^{IF}_{1w1} are the calculated interaction energies, in mJ/m^2, for the particular mineral dispersed in water.*

9.2.1 Generalities; other minerals

Examination of the surface tension components and parameters of the mineral samples listed in Table 9.4 shows that the variability of the γ^{LW} and $\gamma^{\ominus}$ values exceeds those shown in Table 9.1. This can be seen in Figure 9.6 where γ^{LW} is plotted against $\gamma^{\ominus}$. Further, a greater proportion of the minerals in Table 9.4 are hydrophobic compared to the clay minerals in Table 9.1.

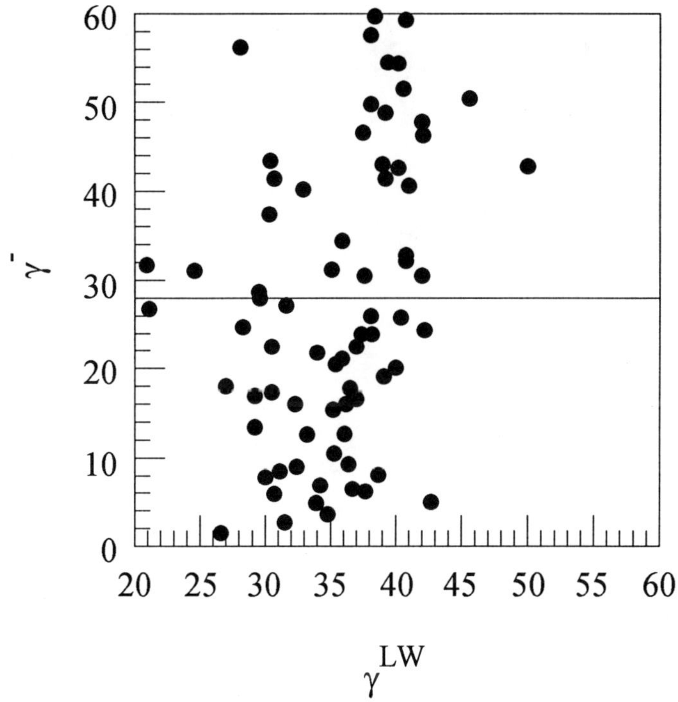

Figure 9.6: *A plot of* $\gamma^{\ominus}$ *versus* γ^{LW} *for the minerals listed in Table 9.4. The horizontal line is the approximate boundary between hydrophilic (above the line) and hydrophobic (below the line) materials.*

10

Biological Interactions with Mineral Particles

10.1 Interactions with Biological Systems

Once introduced into a mammalian biological system, there is no such thing as an *inert* particle. Particles of obvious chemical (covalent) inertness, such as hydrophobic talc particles, can manifest a strong physico-chemical reactivity, in aqueous media, on account of their hydrophobic attraction for biopolymers (e.g., proteins; cf. van Oss *et al.*, 1995b). hydrophilic particles such as silica also bind proteins, by a different physico-chemical mechanism, as do many other metal oxides (van Oss *et al.*, 1995a, b).

The aspecific adsorption of biopolymers, such as serum immunoglobulins, results in the phagocytic ingestion of small particles (Absolom *et al.*, 1982; van Oss 1986). Thus, inhalation by or injection of, e.g., mineral particles in mammals, leads to attempted (if the particles are too large) or successful, phagocytic ingestion by neutrophils and/or monocytes and macrophages, through the intermediary of serum protein adsorption, especially of immunoglobulin-G (IgG)). Adsorption of IgG, as well as of some non-IgG serum proteins such as the most abundant serum protein, human serum albumin, by hydrophobic particles, leads to the activation of human monocytes (monocytes are circulating phagocytic cells which are precursors of macrophages) from peripheral blood (Naim *et al.*, 1998).

Thus, the introduction of small hydrophobic particles (mineral or otherwise) into mammalian systems leads to blood protein adsorption, which in turn leads to phagocytic ingestion (when physically possible) and to the activation of phagocytic cells. For definition and expression of the degree of hydrophobicity, cf. Chapter 9. Particle hydrophobicity is especially important for the adsorption and concomitant partial denaturation of the adsorbed protein (e.g., serum albumin), which is a *conditio sine qua non* for the activation of phagocytic cells (Naim *et al.*, 1998). For the adsorption of IgG on the other hand even adsorption onto hydrophilic, as well as onto hydrophobic particles, can lead to phagocytic ingestion.

10.2 Polymer Adsorption

Most polymers and biopolymers readily adsorb to clay particles from aqueous solution, see, e.g., (Theng, 1979) (general), (Parfitt and Greenland, 1970b), for the adsorption of poly(ethylene glycol) (more usually named polyethylene oxide) onto a Wyoming montmorillonite.

A number of authors have been interested in the adsorption of neutral polysaccharides (dextran and related polyglucosides), e.g., (Parfitt and Greenland, 1970a; Olness and Clapp, 1973, 1975). The adsorption of neutral polysaccharides onto kaolinite as well as onto montmorillonite has been studied (Chenu *et al.*, 1987).

In soils and clays the most generally present biopolymers that naturally occur in the adsorbed state on mineral and clay particles are "humic substances", or "humic acids;" these are decomposition products of lignin, which is the major non-cellulosic polymer in wood and other plant debris. Humic acids (also called "allomelanins" (Merck Index, 1989)) are for the greater part polyphenolic compounds, usually anionic polyelectrolytes, which can complex metal ions, and are surface active and thus capable, upon adsorption onto mineral particles, to enhance their suspension stability in aqueous media (Chheda and Grasso, 1994).

10.3 Protein Adsorption

For the interaction of clay and other mineral particles with and in biological, and especially mammalian systems, the propensity of these particles to adsorb proteins is one of the most important factors. Hydrophobic as well as hydrophilic clay and other mineral particles adsorb proteins from aqueous solutions. However, the mechanisms of protein adsorption in these two cases are entirely different.

10.3.1 Protein adsorption onto hydrophobic surfaces

The adsorption of proteins from aqueous solutions onto hydrophobic surfaces has a simple and straightforward physico-chemical mechanism, based upon the global hydrophobic attraction between an apolar, or largely apolar surface and a hydrophilic polymer (such as a protein), driven by the polar free energy of cohesion of the surrounding water molecules; cf. Chapter 9, section 9.1. Thus, the attraction between a protein molecule and a hydrophobic surface, immersed in water is appropriately designated as a *macroscopic* interaction.

However, subsequent to the initial hydrophobic attraction between a dissolved protein molecule and a hydrophobic surface, a reorganization of the ternary configuration is favored such that the more hydrophobic peptide chains of the protein (which normally are sheltered inside the interior of the dissolved protein's structure), orient themselves toward the protein's exterior, to make closer contact with the hydrophobic surface it is adsorbed to. This reorientation of the protein molecule, occurring upon hydrophobic adsorption, is tantamount to partial denaturation and causes a reorganization which changes the surface structure of the hydrophilic moieties of the protein molecules situated at the water interface (MacRitchie, 1972; Morrissey and Stromberg, 1974). The structural changes brought about by hydrophobic adsorption give rise to enhanced reactivity with human phagocytic cells, whereas protein adsorption onto hydrophilic surfaces does not (Naim *et al.*, 1998). Nonetheless, a certain, but lesser degree of structural change (especially in the secondary conformation) upon protein adsorption onto hydrophilic mineral surfaces also can occur, cf. (Haynes and Norde, 1995; Buys *et al.*, 1996; Maste *et al.*, 1997; Sheller *et al.*, 1998). With time a small proportion of hydrophobically adsorbed protein becomes virtually irreversibly attached, due to hysteresis (Sheller *et al.*, 1998; van Oss *et al.*, 2001a).

10.3.2 Protein adsorption onto hydrophilic surfaces

Upon surface-thermodynamic analysis of protein adsorption onto hydrophilic surfaces such as silica or glass, based on the known surface properties of the hydrophilic protein as well as of the hydrophilic mineral substratum, one would arrive at the conclusion that the macroscopic-scale interactions in a neutral aqueous medium are so strongly repulsive that adsorption should not occur, cf. van Oss *et al.* (1995a). In spite of this prediction, protein adsorption onto hydrophilic glass, silica, etc. does take place, at neutral pH (MacRitchie, 1972; van Oss *et al.*, 1995b). The mechanism of protein adsorption onto hydrophilic surfaces is quite different from that operative with hydrophobic surfaces. Whereas the latter is due to a global, macroscopic scale attraction over the entire surface, hydrophilic adsorption is a microscopic scale attraction of the protein molecules onto tiny discrete cationic sites that are dispersedly distributed throughout the overall anionic mineral surface (van Oss *et al.*, 1995a). Experimental confirmation of the discrete site plurivalent metal ionic attraction for an anionic peptide moiety of the protein is furnished by the observation that the adsorbed protein can be displaced from its site of attachment by complexing agents which specifically bind to plurivalent metal ions, such as Na_2EDTA (van Oss *et al.*, 1995b), or sodium hexametaphosphate. A metal ion site, embedded in an overall hydrophilic anionic surface causes its immediate vicinity to be hydrophobic. Thus, the microscopic scale attraction between a hydrophobic metal ion site and the hydrophilic anionic peptide moiety of a protein molecule, is due to both an electrostatic and a hydrophobic attraction. Nonetheless, in the interaction between a dissolved protein and a metal oxide surface, the overall (macroscopic) repulsion dominates roughly three to five-fold over the local discrete-site (microscopic) specific attraction. This is readily demonstrated by measurement of the kinetic adsorption rate constant of, e.g., the adsorption of human serum albumin onto silica surfaces. Comparing both the overall macroscopic repulsion energies, averaged over all distances and all orientations (leading to net repulsion) of the protein molecules, with the microscopic attraction energies, averaged over all distances and all favorable orientations (leading to net attraction) of the protein molecules, showed that a net average repulsion energy, $\Delta G_{\text{iwj}} \approx +6.91$ kT prevailed between one protein molecule, i, and a discrete silica surface site, j, immersed in water, w, (where 1 kT $= 4.045 \times 10^{-21}$ Joules, at 20°C). Following von Smoluchowski's theoretical analysis (1917) of this type of encounter, the ratio of improbability to probability (f) of the encounter of an albumin molecule with a silica surface was of the order of f $=$ exp(-6.91) $= 10^{-3}$, corresponding to a kinetic *association* (rate) constant $k_a = 7.25 \times 10^4$ L mol^{-1} sec^{-1} (Docoslis *et al.*, 2000a). This does not at all mean that practically no albumin would ultimately adsorb onto silica. Given that the kinetic *dissociation* constant was found to be $k_d = 3.41 \times 10^{-3}$ sec^{-1}, the equilibrium adsorption constant, $K_{eq} = Tk_a/k_d = 2.13 \times 10^{-7}$ L mol^{-1}. this value was only slightly smaller than the K_{eq} found from Langmuir isotherm distributions. It gives rise to a respectable free energy of adsorption, $\Delta G_{\text{iwj}} = -\ln(2.13 \times 10^7 \times 55.56) = -20.9$ kT. It should be noted that as $K_{eq} = k_a/k_d$, and as both Tk_a and k_d are proportional with von Smoluchowski's improbability/probability factor, f, K_{eq} is independent of these improbability/probability considerations which pertain only to the *kinetics* of adsorption and not to its final outcome.

With time, a proportion of the hydrophilically adsorbed protein also becomes practically irreversibly attached, due to hysteresis (van Oss *et al.*, 2001a).

10.3.3 *In vivo* consequences of protein adsorption onto clay and mineral particles

The *in vivo* effects of protein adsorption onto solid surfaces are hard to follow, because of the Vroman effect (Bamford *et al.*, 1992). The Vroman effect is the phenomenon whereby, when solid surfaces are exposed to mixtures of blood plasma proteins, such surfaces adsorb different proteins sequentially, so that with time, first one, then another, and then a third protein, etc., finds itself predominantly adsorbed

A few salient consequences of the adsorption of, *inter alia*, plasma proteins are outlined here, with a view to their connection with the pulmonary pathogenesis of certain asbestos and other clay and mineral particles, discussed in the following section.

Adsorption of the three most abundant plasma proteins (albumin, fibrinogen, IgG) onto hydrophilic surfaces has no apparent effect on monocyte activation, whereas the adsorption of the same plasma proteins onto an otherwise inert but strongly hydrophobic (e.g., silicone) surface, strongly activates these phagocytic cells (Naim *et al.*, 1998).

Adsorption of the immunoglobulin, IgG, onto hydrophilic as well as onto hydrophobic particles causes such particles, coated with (human) IgG, to be ingested by phagocytic cells as a consequence of the fact that phagocytic cells have surface receptors for the Fc moieties of the subclasses, IgG1 and IgG3, of IgG (Absolom *et al.*, 1982).

Needle-shaped mineral particles such as the amphibole asbestos fibers (crocidolite, amosite, tremolite) and the fibrous zeolite, erionite, having adsorbed IgG, but which have a low dissolution rate under intra-phagosomal liquid conditions, are continuously, but largely unsuccessfully, attacked by phagocytic cells, often for periods of years and even decades (van Oss *et al.*, 1999). This causes the fibers to become coated with a "ferruginous" layer consisting of iron-containing proteins (e.g., ferric hemosiderin), as residues of the iron-containing oxidases and peroxidases which are produce by phagocytic cells with a view to the oxidative destruction of ingested particles, such as microbes (van Oss *et al.*, 1999). Whether the thick ferruginous layers around fibrous mineral particles, embedded in the lung or in the pleural cavity, often during decades, are harmful or, on the contrary, whether they lend a long-term (although in many cases not indefinite) protection against pathological sequelae, is not known (van Oss *et al.*, 1999).

10.4 Pulmonary Pathogenesis

10.4.1 Small, roughly spherical particles

In the normal course of events, the occasional inhalation of a few small (no dimensions much greater than a few μm) clay or other mineral particles causes no harm, as such particles become quickly engulfed by phagocytic cells in the lung (mainly macrophages) and are dissolved by the phagosomes within the phagocytes. Even when the rate of dissolution is exceedingly slow, as in, e.g., hydrophobic talc particles, they are ultimately removed by phagocyte-mediated transport, ending usually in gradual elimination through the upper respiratory system (Singer *et al.*, 1972). The main problem with the inhalation of small particles (as defined above) arises when, e.g., workers or miners are continuously exposed to such particles, over long periods of time. In such cases phagocytic removal cannot keep up with the continuous re-supply and the particle overload can then ultimately cause chronic pulmonary pathology. Silicosis is one example of such an outcome. It should be remarked that, other factors being equal, hydrophobic particles are more dangerous than hydrophilic particles. This is because they bind more protein, they activate leukocytes more strongly and they dissolve more slowly intraphagosomally (van Oss *et al.*, 1999). Now, usually silica particles are, rightly, considered to be hydrophilic. However, when freshly ground, or when particles have otherwise been broken or have undergone diminution, silica and most other mineral particles become hydrophobic (Wu *et al.*, 1996). It was found by Vallyathan *et al.*(1995) that the inhalation of freshly silica leads to increased lung inflammation. Talc particles are among the most hydrophobic ones (Giese *et al.*, 1996) and its continuous use involving mucous surfaces and particularly when inhaled (e.g., by babies), or used intravenously, can give rise to severe respiratory disability (Hollinger, 1990; van Oss *et al.*, 1999).

10.5 Needle-shaped or Fibrous Particles

10.5.1 The most dangerous fibrous particles

Clay and other mineral particles that are needle-shaped or fibrous, with a long-axis length that is significantly greater than about 15 μm will,upon inhalation, elicit engulfment by phagocytic leukocytes, which are however too small to engulf them completely. Thus, needle-shaped or fibrous particles tend to remain embedded in the lung tissue, because phagocytes are unable to transport them elsewhere. Scanning electron micrographs of lung tissue obtained from former asbestos workers upon autopsy clearly show the continuing but vain attempts by macrophages to phagocytize the needle-shaped fibers; the coating with Fe-containing protein is also visible on the clean fiber surface which is not covered by macrophages (Sorling and Langer, 1993; van Oss *et al.*, 1999). Those needle-shaped particles that have, in addition, an extremely slow dissolution rate under intra-phagosomal conditions cannot be slowly digested bit-by-bit, so that these needles tend to remain in the lungs for decades and ultimately cause neoplasms. In this category are the amphibole asbestos varieties, crocidolite (blue asbestos), amosite (brown asbestos), tremolite, as well as the zeolite, erionite. The needles of all of these are rather rigid and sharp so that they, once inhaled, tend to pierce the lung tissue and congregate in the pleural sac where, after one or more decades pleural cancer (malignant mesothelioma) ensues. The lag period is typically 20 to 30 years after first exposure to erionite (Baris, 1987) and on an average about 30 years but in a few cases more than 50 years, after exposure to crocidolite, cf. Kane (1993). Malignant mesothelioma is quite rare, but almost invariably fatal. When the disease occurs in someone who earlier in life had been exposed to an amphibole asbestos variety (usually crocidolite), or to erionite, that exposure may be held to have been the cause of the mesothelioma, even when the exposure had been relatively brief, e.g., a year or less (van Oss *et al.*, 1999). This particular outcome of exposure to asbestos appears to have no connection with whether or not the patient had been a tobacco smoker (Kane, 1993).

10.5.2 The less dangerous fibrous (asbestos) particles

In this category belongs mainly the white phyllosilicate asbestos (chrysotile). Now, the rate of dissolution of chrysotile fibers is about an order of magnitude faster than that of the amphiboles and of erionite (van Oss *et al.*, 1999). Chrysotile thus should disappear fairly quickly after inhalation, which is confirmed by autopsy studies (Roggli, 1990). Contrary to amphibole and erionite exposure (which ultimately mainly leads to malignant mesothelioma, see above), long-term chrysotile exposure can lead to lung cancer (Kane, 1993), especially among smokers (Selikoff *et al.*, 1964). Chrysotile asbestos is by far the most common variety used, e.g., in the construction industry. There is little doubt that occasional exposure to chrysotile is not dangerous (phagocytic cells clear away the fibers fairly quickly, thanks to their relatively quick solubility rate). However, for long-term workers with chrysotile, who undergo continuous exposure to the fibers, thus constantly replenishing the material faster than their phagocytes can break it down, exposure to chrysotile remains dangerous, especially if they are smokers, which according to Selikoff *et al.*(1964) increases the danger 50-fold. It appears therefore much safer (and cheaper) not to remove chrysotile asbestos from buildings, but rather to encapsulate the material with a plastic resin, to safeguard the occupants from even minor exposure, and the asbestos removers from further danger. It is of course still essential to perform a prior analysis of the asbestos in question to ensure that it consists of only chrysotile. This is because some of the chrysotile mined, e.g., in Canada, may contain a small but non-negligible proportion of the amphibole tremolite, which is dangerous, even after relatively brief exposure. For the lack of danger of occasional brief exposure to chrysotile, see also Moss (1995).

The reason why needle-shaped amphibole asbestos fibers typically penetrate into the pleural cavity, whereas chrysotile fibers do not, lies in their different morphology and pliability. Microscopic examination reveals that amphibole asbestos fibers are indeed needle-like, rigid for their small diameter (typically only a few μm), and straight. Chrysotile fibers on the other hand are somewhat thinner, often longer, frequently curved, and even somewhat curly, which makes then less capable of piercing living tissue, so that they are more prone to remain *in situ*, until digested.

10.5.3 A few proposed physical or chemical correlations with the pathogenicity of, e.g., amphiboles that turn out to be erroneous

Judging by the differences in iron-contents of crocidolite and chrysotile, Hardy and Aust (1995) proposed that the high iron content of the amphibole asbestos fiber, crocidolite, was the cause of its much greater carcinogenicity than the chrysotile fiber, which, indeed, contains forty times less iron than crocidolite or amosite. However, the fibrous zeolite erionite, also contains much less iron (about 18 times less iron than the amphiboles, and is also much less tightly bound; cf. van Oss *et al.*(1999). As the mechanisms of pulmonary carcinogenesis of amphibole asbestos needles and of erionite needles are remarkably similar, the differences in the iron contents of amphibole and chrysotile asbestos species appear to be irrelevant.

Another apparent difference between chrysotile and crocidolite has been proposed to lie in their electrical surface (ζ) potential (Light and Wei, 1977a, b). In these studies signs of the ζ-potentials of two chrysotile samples (of unknown geographic origin) turned out to be *positive*, whilst the signs of the ζ-potentials of the amphiboles, crocidolite, amosite and anthophyllite were *negative*. Now, positively charged mineral particles are, on the whole, fairly rare (Giese *et al.*, 1996). In another study (van Oss *et al.*, 1999), upon electrophoretic analysis of six chrysotile samples, it was found however that three different samples (two from Quebec, one from Ontario, Canada) were indeed positively charged, but three other chrysotile samples (two from Arizona, one from Zimbabwe) were negatively charged. The amphiboles that were tested (two crocidolites and one amosite sample) were also all negatively charged. Thus the sign of charge of asbestos particles does not correlate with the pathogenicity of the species.

References

Absolom, D. R., van Oss, C. J., Zingg, W., and Neumann, A. W. (1982) Phagocytosis as a surface phenomenon: Opsonization by aspecific adsorption of IgG as a function of bacterial hydrophobicity: J. Ret. Soc., 31, 59-70.

Adamson, A. W. (1990) *Physical Chemistry of Surfaces*: Wiley-Interscience, New York, 664 pp.

Alty, T. (1924) The cataphoresis of gas bubbles in water: Proc. Roy. Soc. London Ser. A, 106, 315-340.

Alty, T. (1926) The origin of the electrical charges on small particles in water: Proc. Roy. Soc. London Ser. A, 112, 235-251.

Andrade, J. D., Ma, S. M., King, R. N., and Gregonis, D. E. (1979) Contact angles at the solid-water interface: J. Colloid Interface Sci., 72, 488-494.

Ashton, A. G., Dwyer, J., Elliott, I. S., Fitch, F. R., Oin, G., Greenwood, M., and Speakman, J. (1984) The application of fast atom bombardment mass spectrometry (FABMS) to the study of zeolotes: In *Proceedings of the Sixth International Conference*, D. Olson and A. Bisio, ed., Butterworths, New York, 704-716.

Bailey, S. W. (1984) *Micas*: Mineralogical Soceity of America, Washington, DC, 584 pp.

Bailey, S. W. (1988a) Chlorites: Structures and Crystal Chemistry: In *Hydrous Phyllosilicates (exclusive of micas)*, Vol 19, S. W. Bailey, ed., Mineralogical Society of America, Washington, DC, 347-403.

Bailey, S. W. (1988b) *Hydrous Phyllosilicates (exclusive of micas)*: Mineralogical Society of America, Washington, DC, 725 pp.

Ball, V., Bentaleb, A., Hemmerle, J., Voegel, J.-C., and Schaaf, P. (1996) Dynamic aspects of protein adsorption onto titanium surfaces: Mechanism of desorption into buffer and release in the presence of proteins in the bulk: Langmuir, 12, 1614-1621.

Bamford, C. H., Cooper, S. L., and Tsuruta, T. (1992) *The Vroman Effect*: VSP, Utrecht, 191 pp.

Bangham, D. H. and Razouk, R. I. (1937) Adsorption and wettability of solid surfaces: Trans. Farady Soc., 35, 1459-1463.

Baris, Y. I. (1987) *Asbestos and Erionite Related Chest Diseases*: Dept. of Chest Diseases, Haceteppe University School of Medicine, Ankara, Turkey, 169 pp.

Bish, D. L., Wu, W., Carey, J. W., Costanzo, P. M., Giese, R. F., Earl, W., and van Oss, C. J. (1999) Effects of steam on the surface properties of Na-smectite: In *Clays for our future. Proc. 11th Int. Clay Conf., Ottawa, Canada, 1997*, H. Kodama, A. R. Mermut and J. K. Torrance, ed., ICC97 Organizing Committee, Ottawa, Canada, 569-576.

Booth, F. (1948) Theory of electrokinetic effects: Nature, 161, 83-86.

Booth, F. (1950) The cataphoresis of spherical, solid non-conducting particles in a symmetrical electrolyte: Proc. Roy. Soc. London Ser. A, 203, 514-533.

Boumans, A. A. (1957) *Streaming Currents in Turbulent Flows and Metal Capillaries*: Ph.D. Thesis, Utrecht, 51 pp.

Bragg, L. and Claringbull, G. F. (1965) *Crystal Structures of Minerals*: G. Bell and Sons, Ltd., London, 409 pp.

Bragg, W. L. (1914a) The analysis of crystals by the X-ray spectrometer: Proc. Roy. Soc. London Ser. A, 89, 468-489.

Bragg, W. L. (1914b) The structure of some crystals as indicated by their diffraction of X-rays: Proc. Roy. Soc. London Ser. A, 89, 248-277.

Brindley, G. W. and Brown, G. (1980) *Crystal Structures of Clay Minerals and Their X-ray Diffraction Identification*: Mineralogical Society, London, 495 pp.

Burst, J. F. (1959) Post diagenetic clay mineral-environmental relationships in the Gulf Coast Eocene in clays and clay minerals: Clays & Clay Minerals, 6, 327-341.

Buys, J., Norde, W., and Lichtenbelt, J. W. T. (1996) Changes in the secondary structure of adsorbed IgG and F(ab')2 studied by FTIR spectroscopy: Langmuir, 12, 1605-1613.

Caillère, S., Hénin, S., and Rautureau, M. (1982) *Mineralogie des Argiles*: Masson, Paris, 184 pp.

Casimir, H. B. C. and Polder, D. (1948) The influence of retardation on the London-van der Waals forces: Phys. Rev., 73, 360-372.

Cassie, A. B. D. (1948) Contact angles: Discuss. Faraday Soc., 3, 11-16.

Chan, D. Y. C., Mitchell, D. J., Ninham, B. W., and Pailthorpe, B. A. (1979) Solvent structure and hydrophobic solutions: In *Water*, Vol 6, F. Franks, ed., Plenum, New York, 239-278.

Chaudhury, M. K. (1984) *Short Range and Long Range Forces in Colloidal and Macroscopic Systems*: Ph.D. Thesis, SUNY at Buffalo, 215 pp.

Chenu, C., Pons, C. H., and Robert, M. (1987) Interaction of kaolinite and montmorillonite with neutral polysaccharides: In *Proc. Int. Clay Conf., Denver*, L. G. Schultz, H. van Olphen and F. A. Mumpton, ed., The Clay Minerals Society, Bloomington, IN, 375-381.

Chheda, P. and Grasso, D. (1994) Surface thermodynamics of ozone-induced particle destabilization: Langmuir, 10, 1044-1053.

Christenson, H. K. (1992) The long-range attraction between macroscopic hydrophobic surfaces: In *Modern Approaches to Wettability*, M. E. Schrader and G. I. Loeb, ed., Plenum, New York, 451.

Clews, F. H. (1971) Clay and ceramic products: In *Materials and Technology*, Vol 2, T. J. W. van Thoor, K. Dijkhoff, J. H. Fearon, C. J. van Oss, H. G. Roebersen and E. G. Stanford, ed., Longman, London, 207-330.

Costanzo, P. M., Wu, W., Giese, R. F., and van Oss, C. J. (1995) Comparison between direct contact angle measurements and thin layer wicking on synthetic monosized cuboid hematite particles: Langmuir, 11, 1827-1830.

Couture, R. A. (1985) Steam rapidly reduces the swelling capacity of bentonite: Nature, 318, 50-52.

Davis, J. A. and Hayes, K. F. (1986) *Geochemical Processes at Mineral Surfaces*: American Chemical Society, Washington, DC, 683 pp.

Debye, P. (1920) van der Waals' cohesion forces: Phys. Zeit., 21, 178-187.

Debye, P. (1921) Molecular forces and their electrical interpretation: Phys. Zeit., 22, 302-308.

Deer, W. A., Howie, R. A., and Zussman, J. (1972) *Rock Forming Minerals*: Longman, London, 333 pp.

de Gennes, P. G. (1990) Dynamics of Wetting: In *Liquids at Interfaces*, J. Charvolin, J. F. Joanny and J. Zinn-Justin, ed., North Holland, Amsterdam, 273-291.

de la Calle, C. and Suquet, H. (1988) Vermiculite: In *Hydrous Phyllosilicates (exclusive of micas)*, Vol 19, S. W. Bailey, ed., Mineralogical Society of America, Washington, DC, 455-496.

Derjaguin, B. V. (1954) A theory of the heterocoagulation, interaction and adhesion of dissimilar particles in solutions of electrolytes: Discuss. Faraday Soc., 18, 85-98.

Derjaguin, B. V. (1989) *Theory of Stability of Colloids and Thin Films*: Consultants Bureau, New York, 255 pp.

Derjaguin, B. V. and Dukhin, S. S. (1974) Nonequilibrium double layer and electrokinetic phenomena: Surface and Colloid Science, 7, 273-356.

Derjaguin, B. V. and Landau, L. D. (1941) Theory of the stability of strongly charged lyophobic soils and the adhesion of strongly charged particles in solutions of electrolytes: Acta Physicoch. USSR, 14, 633-662.

Diao, J. and Fuerstenau, D. W. (1991) Characterization of the wettability of solid particles by film flotation. 2. Theoretical analysis: Colloids Surfaces, 60, 145-160.

Docoslis, A., Giese, R. F., and van Oss, C. J. (2000a) Influence of the water-air interface on the apparent surface tension of aqueous solutions of hydrophilic solutes: Colloids Surfaces B: Biointerfaces, 19, 147-162.

Docoslis, A., Wu, W., Giese, R. F., and van Oss, C. J. (2000b) Measurements of the kinetic constants of protein adsorption onto silica particles: Colloids Surfaces B: Biointerfaces, 13, 83-104.

Docoslis, A., Wu, W., Giese, R. F., and van Oss, C. J. (2001) Influence of macroscopic and microscopic interactions on kinetic rate constants III. Determination of von Smoluchowski's f-factor for HSA adsorption onto various metal oxide microparticles, using the extended DLVO approach: Colloids Surfaces B: Biointerfaces, 22, 205-217.

Doren, A., Lemaitre, J., and Rouxhet, P. G. (1989) Determination of the zeta potential of macroscopic specimens using microelectrophoresis: J. Colloid Interface Sci., 130, 146-156.

Drago, R. S., Vogel, G. C., and Needham, T. E. (1971) A four-parameter equation for predicting enthalpies of adduct formation: J. Am. Chem. Soc., 93, 6014-6026.

Dupré, A. (1869) *Théorie Mécanique de la Chaleur*: Gauthier-Villars, Paris, 484 pp.

Eberl, D. (1984) Clay mineral formation and transformation in rocks and soils: Phil. Trans. Royal Soc., London A, 311, 241-257.

Edberg, J. C., Bronson, P. M., and van Oss, C. J. (1972) The valency of IgM and IgG rabbit aut-dextran antibody as a function of the size of the dextran molecule: Immunochem., 9, 273-288.

Everett, D. H. (1988) *Basic Principles of Colloid Science*: Royal Society of Chemistry, London, 243 pp.

Farmer, V. C. and Russell, J. D. (1971) Interlayer complexes in layer silicates: Trans. Farady Soc., 67, 2737-2749.

Fleischer, M. and Mandarino, J. A. (1995) *Glossary of Mineral Species 1995*: The Mineralogical Record Inc., Tucson, 280 pp.

Fowkes, F. M. (1963) Additivity of intermolecular forces at interfaces. I. Determination of the contribution to surface and interfacial tensions of dispersion forces in various liquids: J. Phys. Chem., 67, 2538-2541.

Fowkes, F. M. (1965) Predicting attractive forces at interfaces. Analogy to solubility parameter: In *Chemistry and Physics at Interfaces*, S. Ross, ed., American Chemical Society Publications, Washington, D.C., 153-167.

Fowkes, F. M. (1983) Acid-base interactions in polymer adhesion: In *Physicochemical Aspects of Polymer Surfaces*, K. L. Mittal, ed., Plenum Press, New York, 583-603.

Fowkes, F. M. (1987) Role of acid-base interfacial bonding in adhesion: J. Adhesion Sci. Technol., 1, 7-27.

Fowkes, F. M., McCarthy, D. C., and Mostafa, M. A. (1980) Contact angles and the equilibrium spreading pressures of liquids on hydrophobic solids: J. Colloid Interface Sci., 78, 200-206.

Franks, F. (1984) *Water*: Royal Society of Chemistry, London, 62 pp.

Fuerstenau, D. W. (1956) Measuring zeta potentials by streaming techniques: Trans. Am. Inst. Min. Engrs., 205, 834-835.

Fuerstenau, D. W., Diao, J., and Williams, M. C. (1991) Characterization of the wettability of solid particles by film flotation. 1. Experimental investigation: Colloids Surfaces, 60, 127-144.

Geertsema-Doornbusch, G. I., van der Mei, H. C., and Busscher, H. J. (1993) Microbial cell surface hydrophobicity: J. Microbiol. Meth., 18, 61-68.

Giese, R. F. (1988) Kaolin minerals: Structures and stabilities: In *Hydrous Phyllosilicates (exclusive of micas)*, Vol 19, S. W. Bailey, ed., Mineralogical Society of America, Washington, DC, 29-66.

Giese, R. F. and Fripiat, J. J. (1979) Water molecule positions, orientations, and motions in the dihydrates of Mg and Na vermiculites: J. Colloid Interface Sci., 71, 441-450.

Giese, R. F., Costanzo, P. M., and van Oss, C. J. (1991) The surface free energies of talc and pyrophyllite: Phys. Chem. Minerals, 17, 611-616.

Giese, R. F., Wu, W., and van Oss, C. J. (1996) Surface and electrokinetic properties of clays and other mineral particles, untreated and treated with organic or inorganic cations: J. Disp. Science and Tech., 17, 527-547.

Girifalco, L. A. and Good, R. J. (1957) A theory for the estimation of surface and interfacial energy. I. Derivation and application to interfacial tension: J. Phys. Chem., 64, 904-909.

Good, R. J. (1967) Physical significance of parameters γ_C , γ_S and Φ , that govern spreading on adsorbed films: SCI Monographs, 25, 328-356.

Good, R. J. (1977) Surface free energy of solids and liquids: J. Colloid Interface Sci., 59, 398-410.

Good, R. J. and Girifalco, L. A. (1960) A theory for estimation of surface and interfacial energies. III. Estimation of surface energies of solids from contact angle data: J. Phys. Chem., 64, 561-565.

Good, R. J. and van Oss, C. J. (1983) Abstr. 57th Colloid Surface Sci. Symp., Toronto.

Good, R. J. and van Oss, C. J. (1991) Surface enthalpy and entropy and the physico-chemical nature of hydrophobic and hydrophilic interactions: J. Disp. Science and Tech., 12, 273-287.

Grasso, D., Smets, B. F., Strevett, K. A., Machinist, B. D., van Oss, C. J., Giese, R. F., and Wu, W. (1996) Impact of physiological state on surface thermodynamics and adhesion of *Pseudomonas aeruginosa*: Environ. Sci. & Tech., 30, 3604-3608.

Gregg, S. J. and Sing, K. S. W. (1982) *Adsorption, Surface Area and Porosity*: Academic Press, New York, 371 pp.

Guggenheim, S. and Martin, R. T. (1995) Definition of clay and clay mineral: Joint report of the AIPEA nomenclature and CMS nomenclature committees: Clays & Clay Minerals, 43, 255-256.

Hamaker, H. C. (1937) The London-van der Waals attractions between spherical particles: Physica, 4, 1058-1072.

Hamilton, W. C. (1974) Measurement of the polar force contribution to adhesive bonding: J. Colloid Interface Sci., 47, 672-681.

Hardy, J. A. and Aust, A. E. (1995) Iron in asbestos: chemistry and carcinogenicity: Chem. Rev., 95, 97-118.

Haynes, C. A. and Norde, W. (1995) Structures and stabilities of adsorbed proteins: J. Colloid Interface Sci., 169, 313-328.

Henry, D. C. (1931) The cataphoresis of suspended particles: Proc. Roy. Soc. London Ser. A, 133, 106-129.

Hiemenz, P. C. (1986) *Principles of Colloid and Surface Science*: Marcel Dekker, New York, 448 pp.

Hollinger, M. A. (1990) Pulmonary toxicity of inhaled and intravenous talc: Tox. Letters, 52, 121-127.

Holmes-Farley, S. R., Reamey, R. H., McCarthy, T. J., Deutch, J., and Whitesides, G. M. (1985) Acid-base behavior of carboxylic acid groups covalently attached at the surface of polyethylene: The usefulness of contact angle in following the ionization of surface functionality: Langmuir, 1, 725-740.

Hower, J., Eslinger, E. V., Hower, M. E., and Perry, E. A. (1976) Mechanism of burial metamorphism of argillaceous sediments: GSA Bull., 87, 725-737.

Hückel, E. (1924) Die Kataphorese der Kugel: Phys. Zeit., 25, 204-210.

Hunter, R. J. (1981) *Zeta Potential in Colloid Science: Principles and Applications*: Oxford University Press, Oxford, 386 pp.

Israelachvili, J. N. (1974) Van der Waals forces in biological systems: Quart. Rev. Biophys., 6, 341-387.

Israelachvili, J. N. (1991) *Intermolecular and Surface Forces*: Academic Press, London, 450 pp.

Israelachvili, J. N. (1992) *Intermolecular and Surface Forces*: Academic Press, New York, 450 pp.

Israelachvili, J. N. and Adams, G. E. (1978) Measurement of forces between two mica surfaces in aqueous electrolyte solutions in the range 0-100 nm: J. Chem. Soc., Faraday Trans. I, 74, 975-1001.

Janczuk, B., Chibowski, E., Bialopiotrowicz, T., Holysz, L., and Kliszcz, A. (1990) Influence of dodecylamine chloride on the surface free energy of kaolinite: Clays & Clay Minerals, 38, 53-56.

Jasper, J. J. (1972) The surface tension of pure liquid compounds: J. Phys. Chem. Ref. Data, 1, 841-1010.

Johns, W. D. and Sen Gupta, P. K. (1967) Vermiculite-alkyl ammonium complexes: Amer. Mineral., 52, 1706-1724.

Kaelble, D. H. (1970) Dispersion-polar surface tension properties of organic solids: J. Adhesion, 2, 66-81.

Kane, A. B. (1993) Epidemiology and pathology of asbestos-related diseases: In *Health Effects of Mineral Dusts*, Vol 28, G. D. Guthrie and B. T. Mossman, ed., Mineralogical Society of America, Washington, DC, 347-359.

Keesom, W. M. (1915) The second virial coefficient for rigid spherical molecules whose mutual attraction is equivalent to that of a quadruplet placed at its center: Proc. Roy. Acad. Sci., Amsterdam, 18, 636-646.

Keesom, W. M. (1920) Quadrupole moments of the oxygen and nitrogen molecules: Proc. Roy. Acad. Sci., Amsterdam, 23, 939-942.

Keesom, W. M. (1921) van der Waals attractive force: Phys. Zeit., 22, 643-644.

Klein, C. and Hurlbut, C. S. (1993) *Manual of Mineralogy*: John Wiley and Sons, Inc., New York, 681 pp.

Korpi, G. K. and de Bruyn, P. L. (1972) Measurement of streaming potentials: J. Colloid Interface Sci., 40, 263-266.

Ku, C. A., Henry, J. D., Siriwardane, R., and Roberts, L. (1985) Particle transfer from a continuous oil to a dispersed water phase: Model particle study: J. Colloid Interface Sci., 106, 377-387.

Kuipers, R. J., Dulfer, R. V., and van der Veen, W. J. (1972) Rubber: In *Materials and Technology*, Vol 5, L. W. Codd, ed., Longman, London, 451-531.

Labib, M. E. and Williams, R. (1984) The use of zeta-potential measurements in organic solvents to determine the donor-acceptor properties of solid surfaces: J. Colloid Interface Sci., 97, 356-366.

Labib, M. E. and Williams, R. (1986) An experimental comparison between the aqueous pH scale and the electron donicity scale: Coll. Polym. Sci., 264, 533-541.

LaFrance, P. J. (1994) *Trajectory modeling of non-Brownian particle flotation using an extended DLVO approach*: MS Thesis, University of Connecticut, Storrs, 55 pp.

Lagaly, G. and Weiss, A. (1969) Determination of the layer charge in mica-type silicates: Proc. Int. Clay Conf. 1969 Tokyo, 1, 61-80.

Lewis, G. N. (1923) *Valence and the structure of atoms and molecules*: The Chemical Catalog Co., Inc., New York, 172 pp.

Li, Z., Giese, R. F., van Oss, C. J., Yvon, J., and Cases, J. (1993) The surface thermodynamic properties of talc treated with octadecylamine: J. Colloid Interface Sci., 156, 279-284.

Li, Z., Giese, R. F., van Oss, C. J., and Kerch, H. M. (1994) Pore size and surface thermodynamic properties of monolithic colloidal gels: J. Amer. Ceram. Soc., 77, 2220-2222.

Lifshitz, E. M. (1955) Effect of temperature on the molecular attracting forces between condensed bodies: Zh. Eksp. Teor. Fiz., 29, 94-110.

Light, W. G. and Wei, E. T. (1977a) Surface charge and asbestos toxicity: Nature, 265, 537-539.

Light, W. G. and Wei, E. T. (1977b) Surface charge and hemolytic activity of asbestos: Environ. Res., 13, 135-145.

London, F. (1930) Theory and systematics of molecular forces: Zeit. Phys., 63, 245-279.

Low, A. J. (1975) Fertility and Fertilizers: In *Materials and Technology*, Vol 7, L. W. Codd, K. Dijkhoff, J. H. Fearon, C. J. van Oss, H. G. Roebersen and E. G. Stanford, ed., Longman, London, 1-88.

Low, P. F. (1961) Physical chemistry of clay-water interaction: Adv. Agron., 13, 269-327.

Lucassen, J. (1979) The shape of an oil droplet suspended in an aqueous solution with density gradient - a new method for measuring low interfacial tensions: J. Colloid Interface Sci., 70, 355-374.

MacEwan, D. M. C. and Wilson, M. J. (1980) Interlayer and intercalation complexes of clay minerals: In *Crystal Structures of Clay Minerals and their X-ray Identification*, G. W. Brindley and G. Brown, ed., Mineralogical Society, London, 197-248.

MacRitchie, F. (1972) The adsorption of proteins at the solid/liquid interface: J. Colloid Interface Sci., 38, 484-488.

Marks, S. (1971) Lime, Cement and Concrete: In *Materials and Technology*, Vol II, T. J. W. van Thoor, K. Dijkhoff, J. H. Fearon, C. J. van Oss, H. G. Roebersen and E. G. Stanford, ed., Longman, London, 91-114.

Maste, M. C. L., Norde, W., and Visser, A. J. G. W. (1997) Adsorption-induced conformational changes in the serine proteinase savinase: J. Colloid Interface Sci., 196, 224-230.

Mayers, G. L. and van Oss, C. J. (1998) Affinity chromatography: In *Enclyclopedia of Immunology*, 2nd Ed., P. J. Delves and I. M. Roitt, ed., Academic Press, London, 47-49.

McTaggart, H. A. (1914a) The electrification at liquid-gas surfaces: Phil. Mag., 27, 297-314.

McTaggart, H. A. (1914b) The electrification at liquid gas-surfaces: Phil. Mag., 28, 367-378.

McTaggart, H. A. (1922) On the electrification at the boundary between a liquid and a gas: Phil. Mag., 44, 386-395.

Merck Index (1989) 11th Ed.

Michaels, A. S. and Dean, S. W. (1962) Contact-angles relations on silica aquagel surfaces: J. Phys. Chem., 66, 1790-1798.

Michot, L., Yvon, J., Cases, J. M., Zimmermann, J. L., and Baeza, R. (1990) Apparente hyrophobie du talc et affinité de l'azote pour le mineral: Comptes Rendus Acad. Sci., Paris, 310, Série II, 1063-1068.

Mijnlieff, P. F. (1958) *Sedimentation and Diffusion of Colloidal Electrolytes.* Ph.D. Thesis, Utrecht, 144 pp.

Moore, D. M. and Reynolds, R. C. (1997) *X-ray Diffraction and the Identification and Analysis of Clay Minerals*: Oxford University Press, New York, 378 pp.

Morrissey, B. W. and Stromberg, R. R. (1974) The conformation of adsorbed blood proteins by infrared bound fraction measurements: J. Colloid Interface Sci., 46, 152-164.

Moss, M. (1995) The schoolroom asbestos abatement program: a public policy debacle: Environ. Geol., 26, 182-188.

Mould, D. L. and Synge, R. L. M. (1954) Biochem. J., 58, 571.

Naim, J. O., van Oss, C. J., Ippolito, K. M. L., Zang, J. W., Jin, L. P., Fortuna, R., and Buehler, N. A. (1998) In vitro activation of human monocytes by silicones: Colloids Surfaces B: Biointerfaces, 11, 79-86.

Napper, D. (1983) *Polymeric Stabilization of Colloidal Dispersions*: Academic Press, London, 428 pp.

Neumann, A. W., Good, R. J., Hope, C. F., and Sejpal, M. (1974) The equation of state approach to determine surface tensions of low-energy solids from contact angles: J. Colloid Interface Sci., 49, 291-304.

Neumann, A. W., Omenyi, S. N., and van Oss, C. J. (1979) Negative Hamaker coefficients I. particle engulfment or rejection at solidification fronts: Coll. Polym. Sci., 257, 413-419.

Neumann, A. W., Omenyi, S. N., and van Oss, C. J. (1982) Attraction and repulsion of solid particles by solidification fronts. 3. van der Waals interactions: J. Phys. Chem., 86, 1267-1270.

Neumann, A. W., Visser, J., Smith, R. P., Omenyi, S. N., Francis, D. W., Spelt, J. K., Vargha-Butler, E. B., van Oss, C. J., and Absolom, D. R. (1984) The concept of negative Hamaker coefficients. III. Determination of the surface tension of small particles: Powder Tech., 37, 229-244.

Nir, S. (1976) Van der Waals interactions between surfaces of biological interest: Progr. Surface Sci., 8, 1-58.

Norris, J. (1993) *Surface Free Energy of Smectite Clay Minerals*: Ph.D. Thesis, SUNY Buffalo, 197 pp.

Norris, J., Giese, R. F., Costanzo, P. M., and van Oss, C. J. (1993) The surface energies of cation substituted Laponite: Clay Minerals, 28, 1-11.

Ohki, S. (1982) A mechanism of divalent ion-induced phosphatidylserine membrane fusion: Biochim. Biophys. Acta, 689, 1-11.

Olejnik, S., Aylmore, L. A., Posner, A. M., and Quirk, J. P. (1968) Infrared spectra of kaolin mineral-dimethyl sulfoxide complexes: J. Phys. Chem., 72, 241-249.

Olness, A. and Clapp, C. E. (1973) Occurrence of collapsed and expanded crystals in montmorillonite-dextran complexes: Clays & Clay Minerals, 21, 289-293.

Olness, A. and Clapp, C. E. (1975) Influence of polysaccharide structure on dextran adsorption by montmorillonite: Soil Biol. Biochem., 7, 113-118.

Omenyi, S. N. (1978) *Attraction and Repulsion of Particles by Solidifying Melts*: Ph.D. Thesis, University of Toronto, 266 pp.

Omenyi, S. N. and Neumann, A. W. (1976) Thermodynamic aspects of particle engulfment by solidifying melts: J. Appl. Phys., 47, 3956-3962.

Overbeek, J. Th. G. (1943) Theorie der Elektrophorese. Der Relaxationseffekt: Kolloid Beihefte, 54, 287-364.

Overbeek, J. Th. G. (1952) Electrokinetics: In *Colloid Science*, Vol 1, H. R. Kruyt, ed., Elsevier, Amsterdam, 194-244.

Overbeek, J. Th. G. and Wiersema, P. H. (1967) Interpretation of electrophoretic mobilities: In *Electrophoresis*, Vol 2, M. Bier, ed., Academic Press, New York, 1-52.

Owens, J. L. and Wendt, R. C. (1969) Estimation of the surface free energy of polymers: Journal of Applied Polymer Science, 13, 1741-1747.

Parfitt, R. L. and Greenland, D. J. (1970a) Adsorption of polysaccharides by montmorillonite: Soil Sci. Soc. Amer., 34, 862-866.

Parfitt, R. L. and Greenland, D. J. (1970b) Adsorption of water by montmorillonite-poly(ethylene glycol) adsorption products: Clay Minerals, 8, 317-324.

Pashley, R. M. (1981) DLVO and hydration forces between micas surfaces in Li^{1+}, Na^{1+}, K^{1+} and Cs^{1+} electrolyte solutions: A correlation of double-layer and hydration forces with surface cation exchange properties: J. Colloid Interface Sci., 83, 531-546.

Pashley, R. M. and Israelachvili, J. N. (1984a) DLVO and hydration forces between mica surfaces in Mg^{2+}, Ca^{2+}, Sr^{2+} and Ba^{2+} chloride solutions: J. Colloid Interface Sci., 97, 446-455.

Pashley, R. M. and Israelachvili, J. N. (1984b) Molecular layering of water in thin films between mica surfaces and its relation to hydration forces: J. Colloid Interface Sci., 101, 511-531.

Pashley, R. M., McGuiggan, P. M., Ninham, B. W., and Evans, D. F. (1985) Attractive forces between uncharged hydrophobic surfaces: Direct measurements in aqueous solution: Science, 229, 1088-1089.

Perry, E. A. and Hower, J. (1970) Burial diagenesis in Gulf Coast pelitic sediments: Clays & Clay Minerals, 18, 165-177.

Powers, M. C. (1967) Fluid release mechanisms in compacting marine mudrocks and their importance in oil exploration: AAPG Bull., 51, 1240-1254.

Pretorius, V., Hopkins, B. J., and Schieke, J. D. (1974) A new concept for high-speed liquid chromatography: Journal of Chromatography, 99, 23-30.

Princen, H. M., Zia, I. Y. Z., and Mason, S. G. (1967) Measurement of interfacial tension from the shape of a rotating drop: J. Colloid Interface Sci., 23, 99-107.

Putnis, A. (1992) *Introduction to Mineral Sciences*: Cambridge University Press, Cambridge, 457 pp.

Reynolds, R. C. (1965) X-ray study of an ethylene glycol-montmorillonite complex: Amer. Mineral., 50, 990-1001.

Rietveld, H. M. (1969) A profile refinement method for nuclear and magnetic structures: J. Appl. Cryst., 22, 65-67.

Rivière, J. C. and Myhra, S. (1998) *Handbook of Surface and Interface Analysis: Methods for Problem Solving*: Marcel Dekker, New York, 976 pp.

Robertson, R. H. S. (1986) *Fuller's Earth - A History*: Volturna Press, Hythe, Kent, UK, 421 pp.

Roggli, V. (1990) Human disease consequences of fiber exposure - A review of human lung pathology and fiber burden data: Env. Health Persp., 88, 295-304.

Rutgers, A. J. and de Smet, M. (1947) Electrosmosis, streaming potentials and surface conductance: Trans. Farady Soc., 43, 102-111.

Rytov, S. M. (1959) *Theory of Electric Fluctuations and Thermal Radiation*: Electronics Research Directorate, AFCRL, Bedford, MA, 265 pp.

Schofield, R. K. and Samson, H. R. (1954) Flocculation of kaolinite due to attraction of oppositely charged crystal faces: Discuss. Faraday Soc., 18, 135-145.

Scott, E. I. and Witney, A. J. (1972) Varnishes, Paints and other Painting Compositions: In *Materials and Technology*, Vol 5, L. W. Codd, K. Dijkhoff, J. H. Fearon, C. J. van Oss, H. G. Roebersen and E. G. Stanford, ed., Longman, London, 317-381.

Seaman, G. V. F. and Brooks, D. E. (1979) Analytical cell electrophoresis: In *Electrokinetic Separation Methods*, P. G. Righetti, C. J. van Oss and J. W. Vanderhoff, ed., Elsevier, Amsterdam, 95-110.

Selikoff, I. J., Churg, J., and Hammond, E. C. (1964) Asbestos exposure and neoplasia: J. Med. Assoc., 188, 22-26.

Shaw, D. J. (1969) *Electrophoresis*: Academic Press, New York, 144 pp.

Sheller, N. B., Petrash, S., Foster, D. M., and Tsukruk, V. V. (1998) Atomic force microscopy and X-ray reflectivity studies of albumin adsorbed onto self-assembled monolayers of hexadecyltrichlorosilane: Langmuir, 14, 4535-4544.

Singer, J. M., Adlersberg, L., and Sadek, M. (1972) Long-term observation of intraveneously injected colloidal gold in mice: J. Ret. Soc., 12, 561-589.

Singh, S. (1997) Packing them in: New Scientist, 28 June, 30-33.

Slade, P. G. and Stone, P. A. (1983) Structure of a vermiculite-aniline intercalate: Clays & Clay Minerals, 31, 200-206.

Slade, P. G., Stone, P. A., and Radoslovich, E. W. (1985) Interlayer structures of the two-layer hydrates of Na- and Ca-vermiculites: Clays & Clay Minerals, 33, 51-61.

Slade, P. G., Dean, C., Schultz, P. K., and Self, P. G. (1987) Crystal structure of a vermiculite-anilinium intercalate: Clays & Clay Minerals, 35, 177-188.

Smith, M. (1971) Natural Graphite: In *Materials and Technology*, Vol II, T. J. W. van Thoor, K. Dijkhoff, J. H. Fearon, C. J. van Oss, H. G. Roebersen and E. G. Stanford, ed., Longman, London, 413-437.

Smyth, J. R. and Bish, D. L. (1988) *Crystal Structures and Cation Sites of the Rock-Forming Minerals*: Allen & Unwin, Boston, 332 pp.

Somasundaran, P. (1968) Zeta potential of apatite in aqueous solutions and its change during equilibration: J. Colloid Interface Sci., 27, 659-666.

Somasundaran, P. (1972) Foam separation methods: Separ. Purif. Meth., 1, 117-198.

Somasundaran, P. and Kulkarni, R. D. (1973) A new streaming potential apparatus and study of temperature effects using it: J. Colloid Interface Sci., 45, 591-600.

Sorling, A. F. and Langer, C. J. (1993) A ferruginous body: New Engl. J. Med., 328, 1388.

Sparnaay, M. J. (1972) *The Electrical Double Layer*: Pergamon Press, Oxford, 415 pp.

Spaull, A. J. B. (1971) Adsorptive Materials: In *Materials and Technology*, Vol 2, T. J. W. van Thoor, K. Dijkhoff, J. H. Fearon, C. J. van Oss, H. G. Roebersen and E. G. Stanford, ed., Longman, London, 185-206.

Srodon, J. and Eberl, D. D. (1984) Illite: In *Micas*, Vol 13, S. W. Bailey, ed., Mineralogical Society of America, Washington, D. C., 584.

Synge, R. L. M. and Tiselius, A. (1950) Biochem. J., 46, xii.

Tabor, D. and Winterton, R. H. S. (1969) The direct measurement of normal and retarded van der Waals forces: Proc. Roy. Soc. London Ser. A, 312, 435-450.

Tanford, C. (1979) Interfacial free energy and the hydrophobic effect: Proc. Natl. Acad. Sci., 76, 4175-4176.

Tanford, C. (1980) *The Hydrophobic Effect*: Wiley/Interscience, New York, 230 pp.

Theng, B. K. G. (1979) *Formation and Properties of Clay-Polymer Complexes*: Elsevier, Amsterdam, 362 pp.

Tuman, V. S. (1963) Streaming potentials at very high differential pressures: J. Appl. Phys., 34, 2014-2019.

Usui, S. and Sasaki, H. (1978) Zeta potential measurements of bubbles in aqueous surfactant solutions: J. Colloid Interface Sci., 65, 36-45.

Vallyathan, V., Castranova, V., Pack, D., Leonard, S., Shumaker, J., Hubbs, A. F., Shoemaker, D. A., Ramsey, D. M., Pretty, J. R., Mclaurin, J. L., Khan, A., and Teass, A. (1995) Freshly fractured quartz inhalation leads to enhanced lung injury and inflammation: Potential role of free radicals: Am. J. Resp. Crit. Care Med., 152, 1003-1009.

van der Waals, J. D. (1873) *Over de continiteit van den gas-en vloestoftoestand*: Ph.D. Thesis, University of Leiden, 250 pp.

van der Waals, J. D. (1881) *Die Kontinuitat des gasformigen und flussigen Zustande*: Barth, Leipzig, 168 pp.

van Olphen, H. (1977) *Introduction to Clay Colloid Chemistry*: Wiley - Interscience, New York, 318 pp.

van Olphen, H. and Fripiat, J. J. (1979) *Data Handbook for Clay Materials and other Non-Metallic Minerals*: Pergamon Press, New York, 346 pp.

van Oss, C. J. (1963) Ultrafiltration membrane performance: Science, 139, 1123-1125.

van Oss, C. J. (1975) The influence of the size and shape of molecules and particles on their electrophoretic mobility: Separ. Purif. Meth., 4, 167-188.

van Oss, C. J. (1979) Electrokinetic separation methods: Separ. Purif. Meth., 8, 119-198.

van Oss, C. J. (1986) Phagocytosis: an overview: Meth. Enzymol., 132, 3-15.

van Oss, C. J. (1994) *Interfacial Forces in Aqueous Media*: Marcel Dekker, New York, 440 pp.

van Oss, C. J. (1995) Hydrophobicity of biosurfaces - origin, quantitative determination and interaction energies: Colloids Surfaces B: Biointerfaces, 5, 91-110.

van Oss, C. J. (1997) Kinetics and energetics of specific intermolecular interactions: J. Molec. Recog., 10, 203-218.

van Oss, C. J. and Beyrard, N. (1963) Effet d'un agitateur magnetique sur las retention des ions dans un ultrafiltre: J. chim. phys. physico. biol., 60, 451-453.

van Oss, C. J. and Fike, R. M. (1979) Simplified cell microelectrophoresis with uniform electroosmotic backflow: In *Electrokinetic Separation Methods*, P. G. Righetti, C. J. van Oss and J. W. Vanderhoff, ed., Elsevier, Amsterdam, 111-119.

van Oss, C. J. and Giese, R. F. (1995) The hydrophobicity and hydrophilicity of clay minerals: Clays & Clay Minerals, 43, 474-477.

van Oss, C. J. and Good, R. J. (1984) The "equilibrium distance" between two bodies immersed in a liquid: Colloids Surfaces, 8, 373-381.

van Oss, C. J. and Good, R. J. (1988) Orientation of the water molecules of hydration of human serum albumin: J. Protein Chem., 7, 179-183.

van Oss, C. J. and Good, R. J. (1990) Estimation of the polar surface tension parameters of glycerol and formamide, for use in contact angle measurements of polar solids: J. Disp. Science and Tech., 11, 75-81.

van Oss, C. J. and Good, R. J. (1991) Surface enthalpy and entropy and the physical chemistry of hydrophobic and hydrophilic interactions: J. Disp. Science and Tech., 12, 273-288.

van Oss, C. J. and Good, R. J. (1992) Prediction of the solubility of polar polymers by means of interfacial tension combining rules: Langmuir, 8, 2877-2879.

van Oss, C. J. and Good, R. J. (1996) Hydrogen bonding, interfacial tension and the aqueous solubility of organic compounds: J. Disp. Science and Tech., 17, 433-449.

van Oss, C. J., Beyrard, N. R., de Mende, S., and Bonnemay, M. (1959) On the separation of ionic isotopes by electrophoresis in gels with pores of the same order of magnitude as the hydrated ions: Comptes Rendus Acad. Sci., Paris, 248, 223-224.

van Oss, C. J., Fike, R., Good, R. J., and Reinig, J. M. (1974) Cell microelectrophoresis simplified by the reduction and uniformization of the electroosmotic backflow: Analyt. Biochem., 60, 242-251.

van Oss, C. J., Gillman, C. F., and Neumann, A. W. (1975) *Phagocytic Engulfment and Cell Adhesiveness*: Marcel Dekker, New York, 302 pp.

van Oss, C. J., Omenyi, S. N., and Neumann, A. W. (1979) Negative Hamaker coefficients II. Phase separation of polymer solutions: Coll. Polym. Sci., 257, 737-744.

van Oss, C. J., Good, R. J., and Chaudhury, M. K. (1986) Solubility of proteins: J. Protein Chem., 5, 385-405.

van Oss, C. J., Chaudhury, M. K., and Good, R. J. (1987a) Monopolar surfaces: Ad. Colloid Interface Sci., 28, 35-64.

van Oss, C. J., Good, R. J., and Chaudhury, M. K. (1987b) Determination of the hydrophobic interaction energy - Applications to separation processes: Sep. Sci. Technol., 22, 1-24.

van Oss, C. J., Roberts, M. J., Good, R. J., and Chaudhury, M. K. (1987c) Determination of the apolar components of the surface tension of water by contact angle measurements on gels: Colloids Surfaces, 23, 369-373.

van Oss, C. J., Chaudhury, M. K., and Good, R. J. (1988a) Interfacial Lifshitz-van der Waals and polar interactions in macroscopic systems: Chem. Rev., 88, 927-941.

van Oss, C. J., Good, R. J., and Chaudhury, M. K. (1988b) Additive and nonadditive surface tension components and the interpretation of contact angles: Langmuir, 4, 884-891.

van Oss, C. J., Chaudhury, M. K., and Good, R. J. (1989a) The mechanism of phase separation of polymers in organic media - Apolar and polar systems: Sep. Sci. Technol., 24, 15-30.

van Oss, C. J., Ju, L., Chaudhury, M. K., and Good, R. J. (1989b) Estimation of the polar parameters of the surface tension of liquids by contact angle measurements on gels: J. Colloid Interface Sci., 128, 313-319.

van Oss, C. J., Giese, R. F., and Costanzo, P. M. (1990a) DLVO and non-DLVO interactions in hectorite: Clays & Clay Minerals, 38, 151-159.

van Oss, C. J., Good, R. J., and Busscher, H. J. (1990b) Estimation of the polar surface tension parameters of glycerol and formamide, for use in contact angle measurements on polar solids: J. Disp. Science and Tech., 11, 75-81.

van Oss, C. J., Giese, R. F., Li, Z., Murphy, K., Norris, J., Chaudhury, M. K., and Good, R. J. (1992a) Determination of contact angles and pore sizes of porous media by column and thin layer wicking: J. Adhesion Sci. Technol., 6, 413-428.

van Oss, C. J., Giese, R. F., and Norris, J. (1992b) Interaction between advancing ice fronts and erythrocytes: Cell Bioph., 20, 253-261.

van Oss, C. J., Giese, R. F., Wentzek, R., Norris, J., and Chuvilin, E. M. (1992c) Surface tension parameters of ice obtained from contact angle data and from positive and negative particle adhesion to advancing freezing fronts: J. Adhesion Sci. Technol., 6, 503-516.

van Oss, C. J., Wu, W., and Giese, R. F. (1993) Measurement of the specific surface area of powders by thin layer wicking: Particulate Sci. Tech., 11, 193-198.

van Oss, C. J., Wu, W., and Giese, R. F. (1995a) Macroscopic and microscopic interactions between albumin and hydrophobic surfaces: In *Proteins at Interfaces, ACS Symposium Series 602*, Vol II, T. A. Horbett and J. L. Brash, ed., American Chemical Society, Washington, 80-91.

van Oss, C. J., Wu, W., Giese, R. F., and Naim, J. O. (1995b) Interaction between proteins and inorganic surfaces - adsorption of albumin and its desorption with a complexing agent: Colloids Surfaces B: Biointerfaces, 4, 185-189.

van Oss, C. J., Giese, R. F., and Wu, W. (1997) On the predominant electron donicity of polar solid surfaces: J. Adhesion Sci. Technol., 63, 71-88.

van Oss, C. J., Giese, R. F., and Wu, W. (1998) On the degree to which the contact angle is affected by the adsorption onto a solid surface of vapor molecules originating from the liquid drop: J. Disp. Science and Tech., 19, 1221-1236.

van Oss, C. J., Naim, J. O., Costanzo, P. M., Giese, R. F., Wu, W., and Sorling, A. F. (1999) Impact of different asbestos species and other clay and minerals particles on pulmonary pathogenesis: Clays & Clay Minerals, 47, 697-707.

van Oss, C. J., Docoslis, A., and Giese, R. F. (2001a) Free energies of protein adsorption onto mineral particles from the initial encounter to the onset of hysteresis: Colloids Surfaces B: Biointerfaces, 22, 285-300.

van Oss, C. J., Wu, W., Docoslis, A., and Giese, R. F. (2001b) The interfacial tension with water and the Lewis acid-base surface tension parameters of polar organic liquids derived from their aqueous solubilites: Colloids Surfaces B: Biointerfaces, 20, 87-91.

van Oss, C. J., Giese, R. F., and Good, R. J. (2002) The zero time dynamical interfacial tension: J. Disp. Science and Tech., 23, (in press).

van Oss, J. F. and van Oss, C. J. (1956) *Warenkennis en Technologie*: J. H. De Bussy, Amsterdam, pp.

van Thoor, T. J. W. (1971) Rock-forming minerals and rocks: In *Materials and Technology*, Vol II, T. J. W. van Thoor, K. Dijkhoff, J. H. Fearon, C. J. van Oss, H. G. Roeberson and E. G. Stanford, ed., Longman, London, 1-90.

Verwey, E. J. W. and Overbeek, J. Th. G. (1948) *Theory of the Stability of Lyophobic Colloids*: Elsevier, Amsterdam, 205 pp.

Verwey, E. J. W. and Overbeek, J. Th. G. (1999) *Theory of the stability of lyophobic colloids*: Dover, Mineola, New York, 205 pp.

Vickerman, J. C. (1997) *Surface Analysis - The Principal Techniques*: Wiley, New York, 457 pp.

Visser, J. (1972) On Hamaker constants - A comparison between Hamaker constants and Lifshitz-van der Waals constants: Ad. Colloid Interface Sci., 34, 331-363.

Vonnegut, B. (1942) Rotating bubble method for the determination of surface and interfacial tensions: Rev. Sci. Instr., 13, 6-9.

von Smoluchowski, M. Z. (1917) Versuch einer mathematischen Theorie der Koagulationskinetik kolloider Lösungen: Zeit. Physik. Chemie, 92, 129-168.

von Smoluchowski, M. Z. (1921) In *Handbuch der Electrizitat und des Magnetismus*, Vol II, L. Graetz, ed., Johan Abvrosius Barth, Leipzig, 374-428.

Wall, C. G. (1972) Petroleum refinery processes and petroleum products: In *Materials and Technology*, Vol 4, L. W. Codd, ed., Longman, London, 37-78.

Warren, A. C. (1973) Paper: In *Materials and Technology*, Vol 2, T. J. W. van Thoor, K. Dijkhoff, J. H. Fearon, C. J. van Oss, H. G. Roebersen and E. G. Stanford, ed., Longman, London, 137-203.

Weaver, C. E. and Pollard, L. D. (1973) *The Chemistry of Clay Minerals*: Elsevier, New York, 213 pp.

White, L. R. (1982) Capillary rise in powders: J. Colloid Interface Sci., 90, 536-538.

Wicks, F. J. and O'Hanley, D. S. (1982) Serpentine minerals: Structures and petrology: In *Hydrous Phyllosilicates (exclusive of micas)*, Vol 19, S. W. Bailey, ed., Mineralogical Society of America, Washington, DC, 91-168.

Wiersema, P. H. (1964) *On the Theory of Electrosphoresis*: Ph.D. Thesis, University of Utrecht, 148 pp.

Williams, D. F. (1972) Cosmetics: In *Materials and Technology*, Vol 5, L. W. Codd, ed., Longman, London, 817-853.

Wood, J. and Sharma, R. (1995) How long is the long-range hydrophobic attraction: Langmuir, 11, 4797-4802.

Wu, W. (1994) *Linkage between ζ-potential and Electron Donicity of Charged Polar Surfaces*: Ph.D. Thesis, SUNY/Buffalo, 185 pp.

Wu, W., Giese, R. F., and van Oss, C. J. (1994a) Linkage between zeta-potential and electron donicity of charged polar surfaces 1. Implications for the mechanism of flocculation of particle suspensions with plurivalent counterions: Coll. Surf. A, 89, 241-252.

Wu, W., Giese, R. F., and van Oss, C. J. (1994b) Linkage between zeta-potential and electron donicity of charged polar surfaces. 2. Repeptization of flocculation caused by plurivalent counterions by means of complexing agents: Coll. Surf. A, 89, 253-262.

Wu, W., Giese, R. F., and van Oss, C. J. (1996) Change in surface properties of solids caused by grinding: Powder Tech., 89, 129-132.

Zisman, W. A. (1964) Relation of the equilibrium contact angle to liquid and solid constitution: Advan. Chem. Ser., 43, 1-56.

Index